essential atlas
of biology

BARRON'S

First English-language edition for the United States, Canada, its territories and possessions published in 2003 by Barron's Educational Series, Inc.

English-language edition © Copyright 2003 by Barron's Educational Series, Inc.

Original title of this book in Spanish: *Atlas Básico de Biología*

© Copyright 2002 by Parramón Ediciones, S.A., World Rights

Published by Parramón Ediciones, S.A., Barcelona, Spain

Authors: Parramon's Editorial Team

Illustrations: Parramon's Editorial Team

Text: José Tola and Eva Infiesta

Translation from Spanish: Eric A. Bye

All inquiries should be addressed to:
Barron's Educational Series, Inc.
250 Wireless Boulevard
Hauppauge, New York 11788
http://www.barronseduc.com

International Standard Book Number 0-7641-2422-6

Library of Congress Control Number 2002107384

Printed in Spain
9 8 7 6 5 4 3 2 1

FOREWORD

This atlas of biology gives the reader a wonderful opportunity to learn about the origin of life, its evolution on Earth, and the characteristics of living creatures, along with their distinct forms. It is thus an extremely useful tool for gaining access to the miracle of life as represented by plants and animals; it will allow us not only to enjoy the variety of life forms but also to see how they are part of our planet's ecology and our own sustenance.

The various sections of this book constitute a complete synthesis of biological science. They contain many schematic but accurate engravings and illustrations that show the main characteristics of what the distinct species of plants and animals are like and how they act. The illustrations, which are the crucial feature in this book, are complemented by brief explanations and notes that make it easier to understand the main concepts; there is also an alphabetical index to help in finding any point of interest quickly and easily.

In undertaking the publication of this atlas of biology, our goal is to create a practical and instructional work that is useful and accessible, strictly accurate from a scientific standpoint, and at the same time, enjoyable and clear. We hope the readers will agree that we have achieved these goals.

CONTENTS

INTRODUCTION

BIOLOGY

This science is devoted to the study of life, as indicated by its very name: in Greek, *bios* means life, and *logos* is the word for science. In fact, it was the Greeks who began the study of life in a scientific way.

One question that arose at the outset concerned the makeup of living organisms, and **Empedocles** (492–432 B.C.) was the first one to define what constitutes humans, plants, and animals. He asserted that there were four elements from which all **matter** was created and that were the moving force of life: water, air, earth, and fire.

At first, biology was intimately connected to medicine, since life was considered a human attribute. Thus, around the sixth century A.D., an important medical school was founded on Kos, one of the Aegean Islands; this school produced many scientists. One of its most illustrious representatives was **Hippocrates** (460–370 A.D.), who established the relationship between the illnesses that affect humans with processes that take place in nature. As a result, the study of life came to include that relationship and, later on, the rest of the creatures that live in nature. Hippocrates held that illnesses must be combated by using the curative forces that exist in nature. However, the person considered as the founder of biological science is **Aristotle** (384–322 B.C.), a philosopher and scientist of considerable prestige in a great many fields and whose authority was upheld through almost 2000 years. The Romans, who did not achieve great scientific advances, passed

down Aristotelian science, and all of Western thought was marked by its doctrine until the time in the modern era when people dared to question the tenets that Aristotle had affirmed 2000 years earlier, in spite of proof to the contrary—as with **spontaneous generation**, which had followers up to the nineteenth century. At that time, **Louis Pasteur** succeeded in proving that no living creature springs from nothingness and can come only from a preexisting one.

But in spite of these errors, which were justified at the time, Aristotle laid the foundation for what is now known as biology. He was the first person who attempted to define the concept of **life**, and he established an initial **classification** of living organisms.

Life forms on Earth have changed throughout time. Species evolve or die out, as with the dinosaurs, which disappeared some 60 million years ago.

Plants and animals are living beings made up of cells, tissues, organs, and systems. They are distinguished from nonliving organisms by the fact that they are born, develop, reproduce, and die.

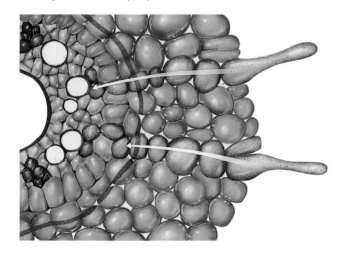

THE FIELDS OF BIOLOGY

A science that focuses on something as essential for us as life necessarily includes many different fields. Throughout the centuries, with the constant increase in acquired knowledge, many of these fields have developed into sciences in their own right, as with **zoology**, which is the study of animals, and **botany**, which focuses on plants. These are natural sciences, devoted to nature, and, consequently, life sciences—in other words, biological sciences.

Other fields seem more distant, or even totally alien to biology. One of these is **physics**. However, knowing and applying to living beings the laws of physics that prevail in the entire universe is very useful in understanding how these organisms work. The converse is also true: It has been very useful to apply to human processes and inventions (airplanes, radar, buildings) the facts that are revealed in living beings. All of that is the focus of **biophysics**.

Chemistry is another of the sciences that has come to include aspects of biology, giving rise to a new

science that is tremendously important in our time, **biochemistry**. Its applications to human health are clear, and, thanks to biochemistry, it is possible to identify the causes and relieve or cure many of the illnesses that plague us, from infections to hereditary or metabolic ones. This is a discipline in which biologists and doctors work together to solve common problems.

Other scientific branches that biologists and doctors study (and which they subsequently put to different uses) are **cytology**, the science devoted to the study of cells, the basic units of life; **histology**, which focuses on the tissues that make up the bodies of plants and animals; and **organology**, the study of tissues organized into larger units known as organs.

The way that all organisms function, whether plant or animal, is the focus of **physiology**. This science investigates the metabolism and the exchange of matter and energy between the organism and the exterior, as well as the ways in which it takes advantage of these resources (which we know as nutrients in the broadest sense) to produce living matter—in other words, how we digest foods and how our body uses them to manufacture its constituent tissues. There are other elements essential to this discipline, such as reproduction and development, the workings of the nervous system and the brain, the relationships of the organism with the outside world through the senses, and any other subject involved in how a plant or animal functions.

One branch of biology that has attained special significance in recent years, but which also has created new problems for humankind, is **genetics**, the science that studies heredity; in other words, the laws

Whereas plants need soil to live and generally stay in one place, animals can move around under their own power. Some birds migrate thousands of miles (kilometers) from bad weather to locate food or shelter.

that govern the transmission of characteristics from one individual to its descendants. One of its applications, **genetic engineering**, is opening up new horizons in medicine and industry that were unimaginable just a few years ago.

Other biological sciences that are entities unto themselves are **ethology**, the study of animal behavior, and **ecology**, which is devoted to all living organisms and the environment they inhabit, the Earth; it focuses on the relationships among organisms and between them and their surroundings.

ONE WAY TO STUDY LIFE

We are going to follow a logical method for examining the materials that biology has to offer. The first questions that arise, as it happened with the ancient Greeks, are what **life** refers to: What is it? Where did it come from? What types of living organisms populate our planet? Then we will see how living beings are distributed over the Earth and examine the relationships they maintain with one another, in other words, an ecological overview.

Once we have examined these general features that encompass the entire field that concerns us, we will move on to a consideration of physical and chemical principles. The first will be the elements that make up **living matter** and the reactions that occur among them. These make up the field of biochemistry and are the basis for explaining many of the topics that will be treated further on. In the final analysis, all creatures from a blue whale down to a tiny one-celled alga are nothing more than molecules of

chemical elements clustered together in different configurations; however, we have to remember that it is possible to construct a skyscraper or a housing development using the same bricks. The difference is in the blueprints.

The following subject will be **evolution**, the passage from the first cell that arose in the primitive sea that covered our planet to the tremendous diversity of organisms that exist today. But in order to explain these changes, we need **genetics** to clarify the laws that make it possible for some species to give rise to others that are more highly evolved. The chemical elements are the building blocks of life, and they can be used to construct the complex forms that represent the various organisms. That will enable us to explain the structure of the plants and animals by studying cells and tissues.

Once we know what living creatures are like, what they look like, and how they are organized, the next step will be to see how they function—in other words, their **physiology**. This is a very complex issue made up of many phenomena, and they have been classified in large systems. One basic one is nutrition; others include transportation of materials inside the organism, movement by means of muscles in the case of animals, and the makeup of the reproductive organs.

There is nothing casual or gratuitous in the world of living beings: everything has its reason for being. Flowers, for example, are brightly colored to attract insects and enlist their aid in pollinating and reproducing.

Once we arrive at that point, we will have an idea of what living beings are like and how they function and behave. Then we will be able to **classify** them in categories according to their similarities. In ancient Greece it was believed that whales were fish because of their hydrodynamic shape and fins and because they live in water. Today we know that they are mammals, thanks to the knowledge about their anatomy and physiology gained since those times.

Next, we will see a general panorama of all living creatures, a type of portrait where each one of them joins others like it in large groups. The old classification of plants and animals has been outdated for many years and instead of those two kingdoms, we speak of five. The **three new kingdoms** are the monera or procaryotes (bacteria and similar organisms that lack a nucleus), the protists (protozoa and one-celled algae with a nucleus), and the fungi.

At that point, we will broaden our perspective with another general overview using the essential knowledge acquired in the preceding steps. We will examine another aspect of **ecology** that involves the relationships among different living beings and with the surroundings in which they live.

This will constitute a complete tour of what has been developed over the centuries as the science of life; this will serve not only to provide the essential knowledge of what the plants and animals that surround us are like, but also to recognize that we are **one more species** on board the great spaceship that is the Earth.

Finally, it must be said that biology is a **composite** science, perhaps more so than the other sciences. This means that a biologist must continually factor in the results of investigations conducted in other fields to avoid losing the sense of interconnectedness and that the laboratory work is worthwhile only if it is adapted to the realities of the organisms that live on our planet.

Mankind, perhaps one of the most recent creatures to appear on Earth, has been a great shaper of the environment.

LIFE: INERT MATTER AND LIVING MATTER

Before we define what life is, we will have a look at a few basic principles on which it is based. We intuitively classify most of the things around us as inanimate objects or living beings, although in some instances it is hard to tell the difference. All of them are composed of atoms, but the way in which they are arranged determines the difference between the ones that are endowed with life and the others that are not.

ATOMS

There are a few more than a hundred **chemical elements** that make up all the living and inanimate matter on Earth and in the universe. The tiniest possible portion of each of these elements is what we refer to as an **atom**. In turn, each atom is made up of **elemental particles** that are common to all atoms; essentially, these are **neutrons, protons,** and **electrons**. The first two join to form the nucleus, and the electrons are arranged in layers and spin around the nucleus, much as man-made satellites spin around the Earth.

THE STRUCTURE OF AN ATOM

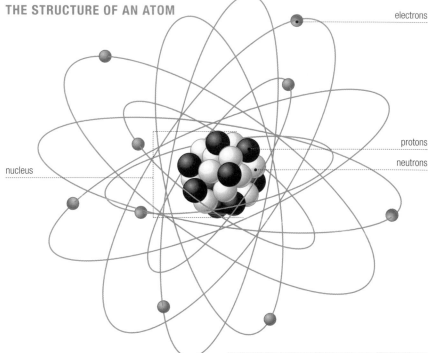

electrons

protons

neutrons

nucleus

So far we know of 105 elements, but it is presumed that there are more—as many as the 118 needed to complete the periodic table.

Quartz crystals.

All elements tend to fill out the number of electrons in each of their layers. Atoms with incomplete layers tend to react with one another.

SOME OF THE MOST COMMON ELEMENTS ON EARTH

Element	Symbol
Hydrogen	H
Sodium	Na
Potassium	K
Chlorine	Cl
Iodine	I
Calcium	Ca
Magnesium	Mg
Sulfur	S
Oxygen	O
Copper	Cu
Iron	Fe
Carbon	C
Silicon	Si
Aluminum	Al
Nitrogen	N
Phosphorus	P

MOLECULES

Atoms have two options for completing their layers of electrons: they can acquire them or share them with other atoms. In the latter case, the atoms that share their electrons join one another to form what is known as a **molecule**. **Water** is one molecule formed by two atoms of hydrogen and one of oxygen. Since oxygen needs two electrons to complete its outer layer (since it has only six electrons and it needs eight), when it joins with two hydrogen atoms it attains the necessary stability.

The union of two atoms that share an electron is called a covalent bond. Hydrogen can form one covalent bond; oxygen can form two.

CHEMICAL TRANSFORMATIONS

Most atoms are joined with others to form molecules, and these molecules, in turn, unite with one another to create new substances. In that way, one or more elements join together to form a larger molecule that is a **chemical compound**. A chemical compound can also split up and give rise to new substances. When all these transformations take place, they consume or release energy. This **energy** is what living creatures use.

Life is possible because of the continuous chemical transformations that occur all over the world.

| CO_2 | $+$ | H_2O | $\Rightarrow$ | H_2CO_3 |
| carbon dioxide | | water | | carbonic acid |

| HCl | $+$ | $NaOH$ | $\Rightarrow$ | $NaCl + H_2O$ |
| acid | | base | | salt   water |

LIVING MATTER

In contrast to what happens with a mineral, which remains unchanged as long as it is not subjected to some manipulation (such as heating, crumbling, or submerging in water), living matter is in a state of **constant change**: It appears (when a plant or animal begins life), then it multiplies (it grows or divides), and eventually it disappears (dies) and is converted to simpler chemical compounds.

Plants are living beings since they develop, breathe, die, and so forth.

Chemical reactions that release energy are called **exothermic.** Chemical reactions that consume energy are called **endothermic**.

THE MOST COMMON ELEMENTS IN LIVING MATTER

Element	Symbol	Percentage
Oxygen	O	62
Carbon	C	20
Hydrogen	H	10
Nitrogen	N	3
Calcium	Ca	2.5
Phosphorus	P	1.14
Chlorine	Cl	0.16
Sulfur	S	0.14
Potassium	K	0.11
Sodium	Na	0.10
Magnesium	Mg	0.07
Iodine	I	0.014
Iron	Fe	0.010
Trace Elements		0.756

WHAT IS LIFE?

It is still difficult to answer this question in spite of the advances produced by science throughout the centuries. The best way to understand something that is *living* is to compare it to something that is inert or not endowed with life. By contrasting them we will understand the basic traits that characterize living beings and distinguish them from inanimate objects.

THE NATURE OF LIFE

One of the most noteworthy characteristics of living beings is that they carry out **functions**. A rock remains immobile through eternity if it is not acted upon from the exterior, but a worm and a bird move, eat, reproduce, die, and serve as food for other creatures. But what about a high-tech robot? Does it move, perform work, feed itself with energy, and even make new robots? Yes, but if we take it apart, we will see that it is made of metal, plastic, screws, and some other mechanical components. Furthermore, it may be made of nothing but aluminum. The **skin** of a worm is made up of thousands of cells grouped in tissues, and every **cell** is made up of many organelles. In addition, the worm and the bird are capable of not only reproducing but of evolving. That is something that robots cannot do.

What differentiates a robot from a simple worm or a beautiful bird is that the latter are endowed with life, and the robot is not.

Granite is a very hard stone, but it is inert; that is, it is incapable of growing or moving by itself.

COMPARISON BETWEEN AN INANIMATE OBJECT AND A LIVING BEING

Characteristic	Inanimate Object	Living Being
Complexity	very low	very high
Functioning	little or none	constant
Perpetuation	no	yes
Irritability	no	yes
Distinct Evolution	no	yes

COMPLEXITY

Living beings are organized on different levels. The most basic level, the atomic level, is similar to the inanimate objects. A second level for chemical compounds is more diverse; so whereas a rock is made up of a dozen compounds (although it could be composed of a single element), a simple amoeba has dozens of them. Above this level, the organisms are more complex and are organized on the basis of **cells**, **tissues** of cells, **organs** of tissues, and **systems** of organs.

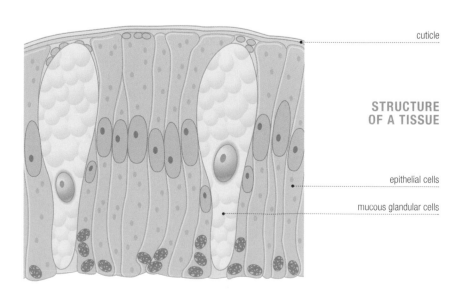

STRUCTURE OF A TISSUE

cuticle

epithelial cells

mucous glandular cells

Introduction

Life

Life on
Earth

The Basis
of Life

Biochemistry

Evolution and
Genetics

Heredity and
Genetics

Functions
of Living
Organisms

How Living
Organisms
Function

Relationships
with the
Outside World

Reproduction
and
Development

Classification of
Living Beings

The Plant
World

The Animal
World

The Living
World

Index

FUNCTIONING

Organisms are constantly exchanging material with the exterior. The process of introducing material to their interior involves **nutrition**; this is followed by the transformation of this material to produce (through **synthesis**) new organic material that will make up the organism's body (its cells, tissues, and so on). But for that to happen, **energy** is needed; this may be sunlight, as with plants, or foods that are rich in energy, as with animals.

A monkey eats fruits produced by a tree with the aid of solar energy.

Metabolism consists of three basic functions: nutrition, respiration, and synthesis of new living material.

PERPETUATION

Living organisms are not eternal, in contrast to inanimate matter. After time, they die; however, life does not disappear, but rather it continues in new organisms. That is, it perpetuates itself. **Self perpetuation** of organisms is possible by means of **reproduction**. This involves producing new organisms at a certain stage of the organism's life; when it dies, they will continue living, and they too will produce other organisms to take their place.

IRRITABILITY

Every living organism, even a lichen clinging to a rock, is capable of reacting to changes in its surroundings, that is, to **stimuli**. This allows them to adapt to the change and survive. This is known as **irritability**, which varies greatly from nearly imperceptible, as in the case of the lichen, to very obvious, as with the **migration** of many birds when winter arrives in the regions they inhabit. Inanimate matter lacks this ability entirely.

One good example of animal irritability is the migration of birds, which spans thousands of miles and is undertaken every year to locate a more favorable environment.

EVOLUTION

The ability to adapt to changes in surroundings makes living beings better or worse off in dealing with their surroundings. They transmit these characteristics to their descendants through units of information known as **genes**. Sometimes there are small changes in the genes; if they constitute an advantage in surviving, they are transmitted to the descendants, which will be slightly different from their ancestors. This slow process, which is referred to as **evolution**, is characteristic of living beings, and it is what makes possible the existence of the diversity that we see today.

THE MOLECULES OF LIFE

All living beings contain the same element, carbon. Although it is also present in inanimate matter, it is characteristic of life. It is present in an infinity of variations and makes up definite structural units, the molecules. All living matter is composed of a small group of molecules combined with one another.

HYDROCARBONS

These are the simplest organic molecules and are made up of only carbon and hydrogen. The simplest one of all is **methane**, whose formula is CH_4. It is produced in nature by the decomposition of animal or vegetable matter, and it also makes up part of the deposits of fuels such as **petroleum** and **natural gas**. Carbon is characterized by its great capacity for bonding with other carbon atoms; that is why there are a great many hydrocarbons made up of a high number of atoms linked together in large molecules.

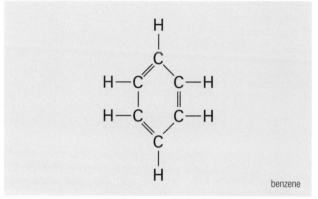

octane

benzene

TYPICAL MAKEUP OF AN ANIMAL CELL

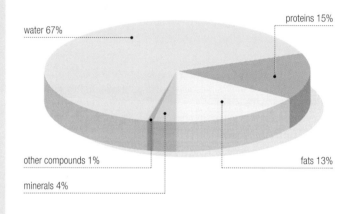

water 67%

proteins 15%

other compounds 1%

fats 13%

minerals 4%

 The formula for glucose is $C_6H_{12}O_6$. Other common sugars are fructose, lactose, and sucrose.

LIPIDS

These compounds are characterized by the fact that they are not soluble in water, or only partially soluble, and that gives them great biological importance. There are three main types of lipids: fats, phospholipids, and steroids. **Fats** are substances that act as an energy reserve for organisms. **Phospholipids** are one of the essential components of cell membranes. **Steroids** are very important substances in animal metabolism; some of the most familiar ones are **cholesterol** and the **estrogens**.

CARBOHYDRATES

These compounds are made up of carbon, hydrogen, and oxygen grouped together in large molecules known as **macromolecules**. There are three main types. **Sugars** are substances that are soluble in water and rich in energy, although less so than fats; however, they can be used more quickly than fats. **Starches** are long chains of glucose and are insoluble in water; as a result, they act as reserve substances. Potatoes and cereals contain large quantities of starch, and that makes them very important foods. Animals also retain glucose in their cells in the form of a special starch named **glucogen**. The third type of carbohydrates is **cellulose**. This is made up of glucose chains that are joined differently than in starches. Cellulose is one of the most abundant organic materials since it makes up the cells and tissues of plants.

Plant fats stored in acorns are food for pigs, which turn them into bacon (animal fat).

PROTEINS

These complex substances are very important to organisms and make up almost half the dry weight of animals' bodies. In addition to carbon, hydrogen, and oxygen, they contain other elements, principally nitrogen and sulfur. Proteins fulfill essential roles for the life of an animal. They make up the **muscle fibers** that make movement possible and intervene in the organism's chemical reactions, acting as catalysts known as **enzymes**.

DNA has four separate nitrogen bases: adenine, guanine, thymine, and cytosine.

A SEGMENT OF DNA (DEOXYRIBONUCLEIC ACID)

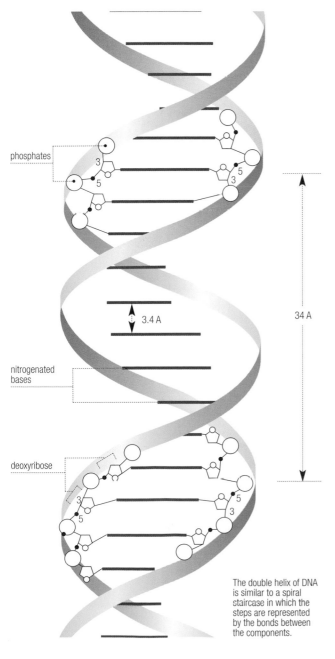

phosphates

3
5

5
3

3.4 A

34 A

nitrogenated bases

deoxyribose

3
5

5
3

The double helix of DNA is similar to a spiral staircase in which the steps are represented by the bonds between the components.

AN AMINO ACID CHAIN

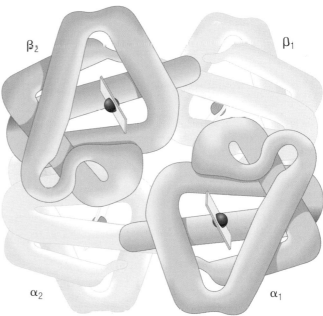

β_2 β_1

α_2 α_1

Somewhat more than 70 **amino acids** are known; they make up all the proteins that exist on Earth; only 20 to 24 of them are found in animals.

Amino acids are the **basic units** that make up proteins; in other words, a protein is a chain (or polymer) of several amino acids.

NUCLEIC ACIDS

These substances are large molecules (**polymers**) made up of a large number of other, simpler molecules know as **nucleotides**. Each one of these is made up of a sugar containing five carbons (a **pentose**), a phosphorous compound (**phosphate group**), and a nitrogen compound (**nitrogenated base**). Nucleic acids facilitate one of the basic functions of life: perpetuation. This is the material that carries **genetic information**. There are two types of nucleic acid: **DNA** and **RNA**.

RNA has four different nitrogenated bases: adenine, guanine, uracil, and cytosine.

THE APPEARANCE OF LIFE ON EARTH

We now know that our planet was formed at the same time as the solar system, around five billion years ago, and more than a billion years had to pass before conditions were right for the first forms of life to appear. There have been a multitude of theories about this phenomenon, and the problem continues to excite the interest of scientists.

MYTHS AND RELIGIONS

For prehistoric people, the nature that surrounded them was frequently threatening, and always a mystery. It was an infinitely superior force, so the power of lightning and the violence of the wind gave rise to the **gods** who controlled these forces. The gods ended up arranged in a complex **mythology**, and they were credited with the creation of the Earth and the appearance of life. Monotheistic religions, such as

Judaism, which worship a single God, endowed the God with all creative powers. The **Bible** compiled all these beliefs and passed them on to Christianity; they have dominated Western thought up to the modern age. All these myths and religions have in common the idea of life's appearance as a willful act on the part of a divinity.

Fragment of *The Creation* by Michelangelo in the Sistine Chapel (Rome).

 Creationism holds that life appeared suddenly by divine will and that it generates itself again after the Flood or any other great catastrophe.

DID LIFE COME FROM OUTER SPACE?

There are theories that affirm that life on Earth arrived from other worlds. Some of them hold that it came from extraterrestrials, but their conclusions have no scientific basis and lack proof. Others indicate that the first **microscopic** creatures that inhabited our planet arrived from space on board **meteorites**. This claim is based on the discovery of organic molecules on some of these meteorites. However, these theories do not explain how life originated and do nothing more than suggest how life might have arrived on Earth.

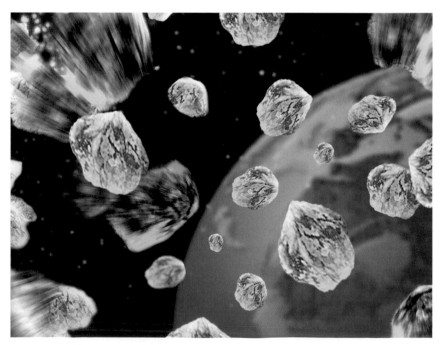

One of the hypotheses on the origin of life on our planet indicates that it arrived on meteorites that landed on Earth billions of years ago.

SPONTANEOUS GENERATION

Because they could not explain how small creatures such as flies suddenly appeared on decaying flesh, the ancients held that they arose spontaneously from the decomposed organic material. This theory was firmly rooted, and although some scientists began to offer proof to the contrary in the seventeenth century, such as the experiments of Francesco Redi, spontaneous generation was not rejected definitively until well into the nineteenth century. Louis Pasteur demonstrated that if no microorganisms from the outside were allowed to reach a culture medium inside a flask, the medium remained sterile; and, if any small creatures appeared, it was because they had been brought there, rather than appearing spontaneously.

Redi's experiment: Flies occurred in the open container with meat but not in the closed one.

Pasteur's experiment: Air can enter through the long tube on the flask but microorganisms cannot; as a result, they do not contaminate the culture medium.

The chaotic conditions on Earth at its inception.

STANLEY MILLER'S EXPERIMENT

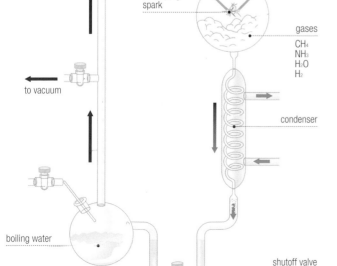

- electrodes
- detonating spark
- gases
  CH₄
  NH₃
  H₂O
  H₂
- to vacuum
- condenser
- boiling water
- shutoff valve

OPARIN'S THEORY

In 1924, Alexander **Oparin**, a Russian biochemist, proposed a theory about the origin of life on our planet. He affirmed that before the Earth's **atmosphere** had any oxygen, organic molecules may have formed in the primitive seas; they floated around in the water and made up a kind of "**primordial soup**." Many of these molecules may have joined with others to form larger ones. At some point, some of these complex structures may have become surrounded by a membrane that separated them from the soup, and when they were subjected to **radiation** from space, they acquired the ability to grow and reproduce, becoming the first living beings.

THE FIRST ORGANIC MOLECULES IN A LABORATORY

In 1952, Stanley Miller decided to reproduce, in the laboratory, the conditions that existed on the planet more than three billion years ago. He mixed **methane** (CH_4), **ammonia** (NH_3), **water vapor**, and **hydrogen** (H_2) in a container in imitation of the primitive atmosphere and circulated it around for several weeks, subjecting it to electric discharges analogous to the lightning bolts from the tremendous storms that occurred at that time. After several weeks had gone by, he was able to identify some **organic molecules** in the condensed water. They were not living beings, but Miller demonstrated that the first molecules that give rise to life could have originated in this way.

LIFE FORMS: THE FIVE KINGDOMS

When life appeared on Earth, it was in the form of tiny, very simple microorganisms. They were clusters of organic molecules united by a membrane and endowed with the ability to feed themselves, grow, and reproduce.

That is when evolution began, and life assumed new forms to take advantage of the available resources. These gave rise to the great number of living creatures we know today, and which we divide into five kingdoms:

THE MONERAN KINGDOM

This kingdom groups together all the simplest organisms known. They are also known as the **prokaryotes** because they have no real nucleus, in contrast to the **eukaryotes**, which include the remaining kingdoms. Prokaryotes have just one **chromosome** for genetic material. Some may have **flagella**, but they are very simple in structure, and they are different from the ones of the protozoa. They reproduce asexually, and only a few of them perform an exchange of genetic material. This kingdom contains two large groups: the **bacteria** and the **blue-green algae**.

PARTS OF A BACILLUS-TYPE BACTERIUM

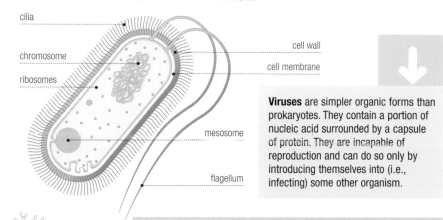

cilia
chromosome
ribosomes
cell wall
cell membrane
mesosome
flagellum

Viruses are simpler organic forms than prokaryotes. They contain a portion of nucleic acid surrounded by a capsule of protein. They are incapable of reproduction and can do so only by introducing themselves into (i.e., infecting) some other organism.

THE PROTIST KINGDOM

This kingdom includes the simplest **eukaryotes**. Their cells contain a true **nucleus** surrounded by a membrane that separates it from the cytoplasm. Most of them are one-celled, and the ones that have more than one cell do not form complex tissues, as plants do. This kingdom includes the organisms known as **protozoa**, which lack chlorophyll, the **algae**, and the slime molds or **myxomycetes**, which are colonial one-celled organisms.

diatom

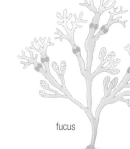
fucus

The algae include the red algae, the dinoflagellates, the euglenophytes, the green algae, the golden algae, and the brown algae.

radiolarian

paramecium

The protozoa include the rhizopods, flagellates, ciliates, and sporozoa.

THE KINGDOM OF FUNGI

These organisms have characteristics partway between the plants and the animals. They have no **chlorophyll**, which means that, like animals, they cannot produce their own food and are **heterotrophs**; however, they reproduce by means of **spores**, like plants. They are made up of filaments known as hyphae. In larger fungi, the hyphae interlace to form a tissue referred to as the **mycelium**. Mushrooms are the part of certain fungi that grow through the topsoil; others are microscopic.

Many fungi are parasites that feed on plants and animals; others produce antibiotics like penicillin, are used to make cheese like Roquefort, and grow mushrooms that are either edible or poisonous.

Three poisonous mushrooms

Amanita muscaria

Amanita phalloides

Coprinus atramentarius

Lichens are associations between fungi and algae.

THE PLANT KINGDOM

This kingdom includes all the **multicelled** organisms that have **chlorophyll** and are therefore capable of producing their own food using water, mineral salts, and energy from the sun; these organisms are **autotrophs**. These are known as **higher plants**, and, in contrast to the lower plants (grouped together in the protist kingdom, such as the algae), they have **differentiated tissues** and reproduce either asexually or sexually.

Most of the plants that we see are higher plants.

The plant kingdom includes the **bryophytes** (mosses and liverworts) and the **vascular plants** (that have conductor vessels).

The vascular plants include **ferns**, **gymnosperms** (conifers), and **angiosperms** (dicotyledons and monocotyledons).

Animals are classified as **invertebrates** (possessing a soft body or an external skeleton) and **vertebrates** (those with an internal skeleton made up of vertebrae and other bones).

THE ANIMAL KINGDOM

This kingdom includes all the **multicelled** organisms that have no chlorophyll, and which, thus, have to feed themselves on existing organic material; they are known as heterotrophs. Some of them eat plant matter and are **phytophagous**; others eat other animals and are **carnivorous**. There are also **parasitic** species. Animals differ from the plants also because their tissues are more complex and they can move about under their own power. Additionally, they are equipped with **nervous tissue**.

In contrast to the plants, animals have to feed on existing organic material.

FROM WATER TO LAND

Life appeared on a naked planet where only inanimate matter existed. It developed on that planet and continues to do so in our time. This matter is the scenario where life takes place, and it can come in many different forms. Its great variety is one of the causes for the existence of such diverse organisms as protozoa, giant squids, and dogs.

THE LIQUID MEDIUM

Life arose in the primitive sea that covered the Earth, and water came to be a major component of living creatures; as a result, some animals, such as a jellyfish, can be up to 99% water. Water is essential to life because of its physical and chemical properties. Its **viscosity** and **density** make it possible for fish, mollusks, and coral to live in it; its high **caloric capacity** is indispensable in regulating environmental temperature changes, its **dielectric constant** is crucial for many reactions, and more.

The ocean continually has a high dielectric constant, which explains the facility with which salts ionize in it.

THE WATER CYCLE: A CONSTANT CIRCULATION THROUGHOUT THE PLANET

rain and snow
glacier
lake
river
underground water

condensation
water vapor
evaporation
ocean

APPROXIMATE COMPOSITION OF FRESHWATER AND SALT WATER
(in grams per liter)

Salt Water		Freshwater	
NaCl	26.52	$Ca(OH)_2$	0.030
$MgSO_4$	3.30	$MgSO_4$	0.020
$MgCl_2$	2.25	NaCl	0.015
$CaCl_2$	1.14	KCl	0.006
KCl	0.72	$CaCl_2$	0.003
$NaHCO_3$	0.20		
NaBr	0.08		

The density of air and the quantity of oxygen decrease with altitude.

Oxygen and carbon cycles. Both elements are in constant circulation through the biosphere.

THE ATMOSPHERE

The planet is surrounded by a **gaseous covering** that is much less dense than water and, therefore, can be inhabited only in the areas that are close to the **ground**. Flying animals, such as insects and birds, use the atmosphere as a medium for movement, but they also need solid ground to carry out much of their activities. Nowadays, the **atmosphere** is different from what existed at the beginning of the Earth's history and that is due to **photosytnthetic** organisms that have enriched it with oxygen. It is precisely this appearance of **oxygen** in high quantities in the atmosphere that made possible the development of land plants and animals.

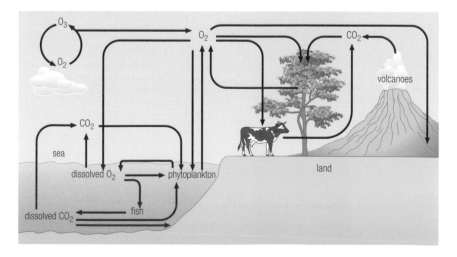

O_3
O_2
O_2
CO_2
CO_2
volcanoes
sea
dissolved O_2
phytoplankton
land
dissolved CO_2
fish

PRESENT COMPOSITION OF THE ATMOSPHERE

Component	%
Nitrogen	78.08
Oxygen	20.94
Argon	0.93
Carbon dioxide	0.03

All life occurs in the lowest layers of the troposphere (close to the ground), which extend upward for about seven miles (11 km).

RADIATION AND ENERGY

Our planet receives light from the **sun** as the main external source of energy, but radiation also comes from more distant regions of the **universe**. There are several types of **radiation**, which is classified according to its **wavelength**. The shorter the wavelength, the more energy it carries and the more penetrating it is, but excessively intense radiation is dangerous and even fatal to living creatures. The **atmosphere** acts as a protective layer that keeps the most dangerous radiation from reaching the ground and allows passage of what we refer to as **visible light**. This type of radiation is the best one for life, since it supplies the right amount of energy for biological processes such as **photosynthesis**.

TYPES OF RADIATION

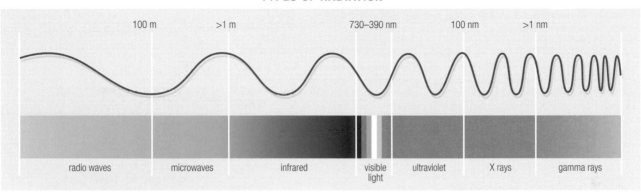

100 m	>1 m	730–390 nm	100 nm	>1 nm
radio waves	microwaves	infrared	visible light	ultraviolet	X rays	gamma rays

PRINCIPAL TYPES OF RADIATION
(1 nm = 0.000001 mm)

	Ultraviolet (short wave)	Visible light	Infrared (long wave)
Wavelength	< 360 nm	360–750 nm	> 750 nm

THE SOIL

The solid substrate on which the animals and plants live is what we commonly call **soil**, but not all substrates are soil. Soil is the result of the activity of **organisms** on the inorganic substrate. In other words, the **atmospheric agents** (water, wind, etc.) erode and wear down the solid rocks, which are broken into smaller rocks, pebbles, and sand. At the same time, the organisms that live on top produce wastes and leave their remains behind when they die. All this **organic** matter (transformed by bacteria, fungi, and so forth) mixes with the **inorganic** matter to produce soil, which is capable of sustaining plant life.

A

B

C

Soil is arranged in layers known as **horizons**. In general terms, the soil is divided into three horizons: **A**, in which organic material is predominant (this is the most superficial layer); **B**, where there is a balance between organic and inorganic matter; and **C**, where most of the material is inorganic and in contact with the underlying rock.

Soil offers plants nutrients and a place to grow, and shelter for many animals (nests, burrows, etc.).

THE BIOSPHERE

Only a small part of our entire planet is suited for life and that is where plants and animals have developed. This area is what is known as the biosphere, and it is where many biological phenomena occur.

These phenomena take place with respect to three components of the Earth: the gaseous element (air) that makes up the atmosphere, the liquid element (water) that constitutes the hydrosphere, and the solid element (rock) that is the lithosphere.

THE BIOSPHERE: OUR HOME

Seen from space, our **planet** looks like a blue ball. That is because of the presence of the **water** and the **atmosphere**, but also because of the existence of **life**. We living beings occupy the lowest layers of the atmosphere, the uppermost layers of the oceans, and barely a few yards (meters) of the Earth's crust; however, we react with all these mediums

in such a way that we create our own environment, which we call the **biosphere**. We depend on it to live, and it, in turn, depends on us, since, if there were no life, it would not take long for the oxygen to disappear from the atmosphere, the composition of the water would change, and erosion would quickly transform the Earth's surface.

ANNUAL PRODUCTION OF VEGETATION IN THE BIOSPHERE

	Surface area (in millons of km^2)	Carbon production (in millons of metric tons)
Forests and jungles	41	16,400
Cultivated lands	15	5,250
Pasture and grasslands	30	6,000
Arid zones and deserts	40	2,000
Oceans	361	36,100
Rivers and lakes	1.9	190
TOTAL	488.9	65,940

The ozone layer is located at an altitude between 18 and 30 miles (30 and 50 km) above the troposphere.

The biosphere is restricted to the troposphere, which reaches up to 7 miles (11 km).

THE ATMOSPHERE AND LIFE

We have already seen how the present atmosphere came about as a result of the activities of living beings. But ever since the amount of available **oxygen** allowed the existence of organisms outside the water, there has been an ongoing change in the relationships of the plants and animals with this atmosphere. Half of the atmospheric mass is concentrated near the ground at an altitude of a little more than 15,000 feet (5,000 meters), and the density decreases after that point. Although the atmosphere is chemically stable, it is a very dynamic environment with displacements (**wind**, **storms**), variations in the quantity of water (**humidity**, **clouds**), and thermal swings. These changes have a major impact on organisms, which have evolved and adapted to them: plants in dry or wet areas, terrestrial animals in deserts or mountains, soaring birds, and so forth.

The atmospheric phenomena taken together constitute the **climate**.

Clouds carry great quantities of water and deposit it on the Earth in the form of rain or snow.

THE HYDROSPHERE

The layer of water that surrounds the planet is known as the **hydrosphere**; it has some characteristics that are comparable to those of the atmosphere but with some important differences. One essential difference is the **density** of water, which is some 770 times greater than that of air. This makes it possible for organisms to spread out in it and use it for support. They have developed special structures for this purpose, such as **fins**, which are present in practically all aquatic animals (fish, marine mammals, etc.). The hydrosphere is another dynamic environment. **Waves** are produced on the surface of the **ocean** as a result of wind and **currents** below the surface, because of differences in density and temperature. **Rivers** are in constant movement from higher areas to the point where they enter the ocean or a **lake**. Lakes share some characteristics with the oceans and rivers.

Waves are produced by the effect of wind on the surface of the sea. When the waves hit the coast, they contribute to erosion and shaping of the coastline.

Even outer space has an influence on ocean dynamics: The sun and moon cause the tides. ←

Animals are forced to adapt to the nature of the substrate. Goats have developed sharp hooves for climbing on bare rock, and camels have broad feet for walking on sand. ↑

It is a paradox that some deserts go right up to the edge of the ocean, but it is the lack of rain and the dryness of the air that cause deserts.

THE LITHOSPHERE

The lithosphere is the solid mass of the planet, but we living creatures use only a small portion of it. The parts that are vital for life are the surface, on which the plants and land animals live, and the layer of **soil**, which generally extends only for a few feet below the surface. Only the roots of some plants penetrate down to a level of several dozen feet in search of water, and the area in which animals live coincides with the fertile layer that fosters plant growth. One exception is **caves**, which can penetrate hundreds of feet into the crust, but the atmosphere, and sometimes even the hydrosphere, gets into them.

LIFE FORMS AND THEIR SURROUNDINGS

Now we will see how the kinds of relationships that exist between living beings and their physical surroundings have created what we commonly refer to as landscapes—in other words, natural environments. These are the result of millions of years of activity involving plants and animals, and for that reason they are dynamic—that is, they are in constant flux.

NATURAL ENVIRONMENTS

For biologists, the **natural environment** is equivalent to the **ecosystem** that we will study later on and that we will define. Right now we are going to look at these environments like the landscapes that geographers study, in other words, by giving the same importance to the physical environment and the living organisms that inhabit it. We are doing it this way to emphasize the great scenarios in which life developed and continues to develop. The first distinction that we can make is between **aquatic** and **dry land** environments. The former include the oceans and all the water that exists on the continents. The land environments are differentiated from one another by topography (flat or mountainous) and by the vegetation that is present (forests or deserts).

The rivers, lakes, swamps, and other continental waters contain only a tiny part of all the water on Earth, but they are tremendously important in an ecological sense.

MOUNTAINS

Mountains are a rough geographical accident that forces animals to make special adaptations, such as the nonsliding hooves of the chamois. Also, they are everywhere characterized by their vegetation, which is distributed in layers. As a result, very different types of vegetation appear concentrated in a small area, ranging from the **jungles** at the base, to the **temperate forests** at an intermediate altitude, the **tundra** of high areas, and an area of **year-round snow** with conditions similar to the polar regions.

perpetual snow
(polar regions)

alpine meadows
(tundra)

conifer forest
(taiga)

deciduous
forest
(temperate
forest)

jungle

The oceans are a mass of salt water that covers more than two-thirds of the surface of our planet.

The oceans cover 70 percent of the surface of the planet and reach a maximum depth of a little more than seven miles (11 km). Plants (mainly **algae**) live only in the upper part, no deeper than light can penetrate, and most of them are attached to the bottom. **Animals**, on the other hand, are found distributed throughout the liquid mass, some attached to the substrate (such as sponges), others moving about on it (like lobsters); others freely swim through it near the **coast**.

POLAR REGIONS AND TUNDRA

At the two ends of the Earth's axis, the sun's rays fall at a greater angle and provide much less heat than at the equator. Temperatures are very low, and ice forms an extensive shell. In the **North Pole** or the **Arctic**, the mass of ice floats on top of the sea; at the **South Pole**, however, the ice covers the continent of **Antarctica**. In land areas nearest the polar regions, the low temperatures allow only lichens, mosses, and some dwarf superior plants to live. This environment is known as the **tundra**.

Because of the severe climate in Antarctica, plant life is limited to some lichens, algae, and moss; and animal life to a few insects, penguins, and seals.

Animals characteristic of the tundra include the arctic fox, the lemming, and the caribou.

DESERTS AND ARID REGIONS

Deserts form in certain areas of the planet because of a lack of water. They may be hot (like the **Sahara**) or cold (like the **Gobi**), and some are sandy, with dunes, and others are stony. Some deserts contain no plant life (the **Atacama**, for example), but in others the rains that fall every couple of years allow **cacti** and similar plants to grow. The transitional area between the deserts and the more humid regions are the **arid** and **semiarid zones**. The type of vegetation in each area depends on the amount of available water, but, generally, the plants are herbaceous, with some scrub brush.

Dunes in the Rub al Khali desert in the Arabian Peninsula.

SAVANNAS AND PRAIRIES

Prairies (such as the **Great Plains** and the Russian **steppes**) and **savannas** (in Africa) exist in flat areas that receive moderate rains. The predominant vegetation is herbaceous plants, especially grasses, with some scrub brush, and sometimes a few trees. Under these conditions, almost all animals are adapted to running.

The savanna characterizes hot regions that have a long dry season; the principal plants are bushes and tall grasses.

The taiga is a forest of conifers in cold regions.

FORESTS AND JUNGLES

When there is plenty of precipitation in both plains and mountains, the climax plants are trees, which cluster together to form **forests** and **jungles**. In temperate regions it is usual to speak of forests; the appropriate term in hot and equatorial regions that have almost continual rain is *jungle*. Forests can be made up of **coniferous** or **deciduous** trees.

Jungles have thick vegetation because of ample rain and a hot environment.

FROM ISOLATED INDIVIDUAL TO POPULATION

Just as organisms react to the environmental conditions in which they live, they also react to the presence of other living beings. Relationships are established among different individuals that may be very different from one another, but there are others that compete with each other or hunt one another.

RELATIONSHIPS AMONG CELLS

The **cell** is the smallest unit of life. In some cases, a single cell can make up a complete living organism; this is the case with the **protozoa**. In other cases, though, cells are part of **tissues**. In that case, an individual cell lives in a very close relationship with other cells of the same kind. This relationship can be of many different types. In a protective tissue such as the **skin**, the cells that make it up are closely linked with one another so that there are no holes that could allow the entry of harmful substances or organisms into the body. In **muscle tissue**, the cells are coordinated so that they relax and contract all together. In the **nervous tissue**, each **neuron** acts as a relay station for the signal that comes in from some distant point and must be transmitted to another place.

The cells of a tissue communicate with one another and exchange nutrients, but they retain their individuality.

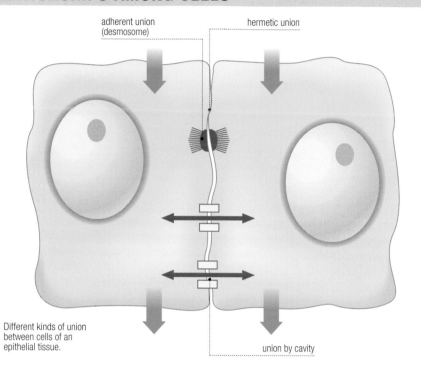

Different kinds of union between cells of an epithelial tissue.

adherent union (desmosome)

hermetic union

union by cavity

RELATIONSHIPS AMONG ONE-CELLED ORGANISMS

Even among one-celled organisms, there are basic relationships of the type that exist among all living beings. On the one hand, there are the organisms that contain chlorophyll and that make their own food; these are known as autotrophs. On the other hand, there are the organisms that must feed themselves organic material that is available from other sources because they do not have chlorophyll; they are heterotrophs. Among the latter there are many ways of obtaining their food, such as hunting other organisms, feeding on dead matter, and parasitizing plants and animals.

AUTOTROPHIC AND HETEROTROPHIC ORGANISMS

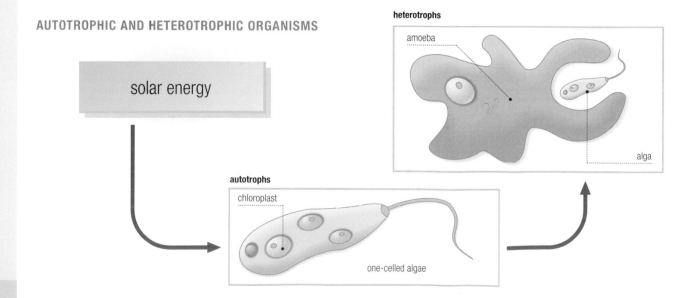

heterotrophs

amoeba

alga

solar energy

autotrophs

chloroplast

one-celled algae

PLANTS AND ANIMALS

These two types of **multicelled** organisms can be very complex. In general, the more highly evolved they are, the greater the number of relationships that they can maintain directly or indirectly with other organisms. Plants commonly compete with one another for available resources, such as water and light. In order to reach these resources, the plants grow faster or higher, they climb on other plants (as **vines** do), and they even produce substances that are harmful to other plants. Their relationships with animals are commonly defensive: they have a hard bark, grow **thorns**, and so forth.

The different animal species have come to inhabit all the environments on the planet: the air, the land, and the waters.

In the dense jungle, the various species of plants and animals compete either to reach more light and more food or to occupy a space where they are protected.

Animals maintain various types of relationships with other organisms:
phytophagous animals feed on plants;
carnivores eat other animals;
scavengers or carrion eaters eat the flesh of dead animals; and
parasites obtain their nutrients from a host that they occupy but do not kill.

POPULATIONS

A population is defined as the set of a specific type of organisms in a given area. We can speak of the population of animals of an island, or the population of mammals on the island, or of a population of bats that live in the caves of that island, even to the point that we identify a population of a species of bat (such as the long-eared bats) in cave 37 of the island. This makes it easier to study that species. This method is frequently used by biologists.

It is easy to see how zebras, gnu, and gazelles occupy the same territory.

The size of a population is very important for a species. If it drops below a certain level, the animal becomes endangered.

THE EARTH, OUR LABORATORY

The Earth is part of the solar system and contains the same elements, although in different proportions, compared to other planets. Most of the elements react with one another to produce compounds.

Except for the first organic matter that gave rise to life, the compounds are inorganic. However, they are of great importance to living beings, and they are part of the chemistry of plants and animals.

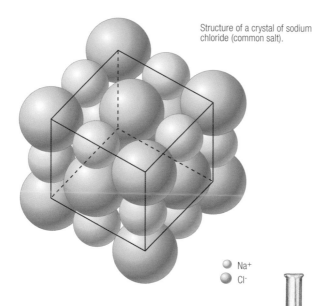

Structure of a crystal of sodium chloride (common salt).

○ Na⁺
○ Cl⁻

CHEMICAL BONDS

Generally, the **chemical elements** do not occur in isolation but are incorporated into **molecules**. The force that holds them together is called a **chemical bond**; this is what makes it possible to use two or more elements to produce a substance with new properties. The three main types of bond are **ionic**, which is the bond established between **ions** (atoms with an electrical charge); **covalent**, which is formed among atoms that have the same affinity for taking on electrons; and **polar covalent**, when two atoms share their electrons.

→ The energy required to break a bond is the same that is required to form it; it is known as the **bond energy**.

HYDROPHILIC AND HYDROPHOBIC SUBSTANCES

There is a classification of molecules into two types that are very important for living beings. This involves substances or molecules that are **hydrophilic**—those that are attracted to water—and those that are **hydrophobic**, or which shun water. This makes possible substances that dissolve in water, and others that are insoluble. Thanks to these characteristics, cells and organisms in general can create impermeable membranes to keep their interior constant.

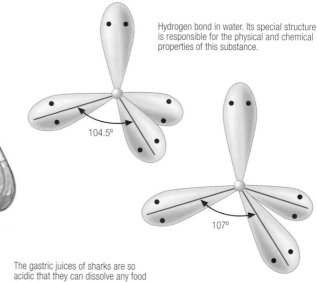

Hydrogen bond in water. Its special structure is responsible for the physical and chemical properties of this substance.

104.5°

107°

↓ The **hydrogen bond** is not as strong as a covalent bond, but it is strong enough to keep the water molecules (chemical formula: H_2O) together. Each one unites with four others because each oxygen atom is associated with two hydrogen atoms. One consequence of these characteristics is the ability of water to remain liquid at mild temperatures. This made the emergence of life possible.

The gastric juices of sharks are so acidic that they can dissolve any food that is ingested.

ACIDS AND BASES

Chemists define an **acid** as a substance that give up protons to another substance; a **base** is a substance that accepts protons from another substance. In other words, the two of them behave in opposite ways. This characteristic makes them react very intensely with other substances. Living beings contain both types, for example the hydrochloric acid in the **gastric juices** of animals and the **bicarbonate** in the tissues of many organisms.

pH

With many ordinary products, we read that the pH is similar to that of skin, so they are harmless to it. What is the pH? It is an indication of the degree of acidity of a substance; it is important because the membranes of many cells are destroyed if the surroundings are too acidic or too basic. In more scientific terms, pH is defined as "the negative logarithm of the concentration of hydrogen ions."

The pH of water is 7; in other words, water is neutral.

THE pH OF WATER

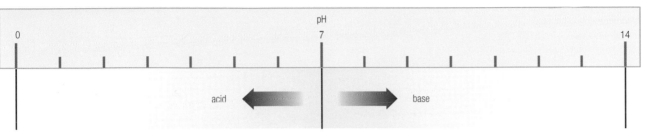

pH

0   7   14

acid   base

1×10^{0} moles/liter   pure water 1×10^{-14} moles/liter

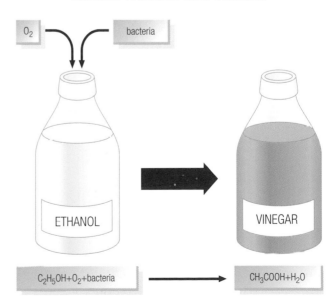

For farmers, the pH of the soil is very important for the types of vegetables they want to raise and for the type of fertilizer they need to use to enrich the soil.

OXIDATION-REDUCTION REACTIONS

We have already seen that chemical bonds involve giving up, taking on, or sharing **electrons**. It is said that an atom that gives up electrons **oxidizes** and that one that takes them on **reduces**. The reactions are thus **oxidizing** and **reducing** reactions, respectively. The most important thing is that they take place at the same time, since, in order for one substance to oxidize, another has to reduce. Therefore, this simultaneous combination of oxidation and reduction is called a **redox reaction** for short.

TURNING ALCOHOL INTO VINEGAR

O_2 bacteria

ETHANOL   VINEGAR

$C_2H_5OH + O_2 + bacteria$ $CH_3COOH + H_2O$

The environmental pH has a significant effect on plant life on our planet. Even though some plants have a high tolerance for acidic water and soil, others react more favorably to alkaline conditions; however, they may be unable to adapt to excessively acidic or alkaline environments.

There are many redox reactions that take place in nature; examples include the transformation of alcohol into acetic acid and water (vinegar) through the action of bacteria.

PLANTS: FOOD FACTORIES

Animals exist thanks to plants, which are the only organisms that are capable of transforming inorganic material from the environment into organic matter with the aid of water and solar energy. This process takes place inside the chloroplasts of the leaves, which function like tiny factories.

LEAVES AND CHLOROPLASTS

Chloroplasts are **flat structures** found inside plant cells, sometimes in very large numbers (up to 50). They are surrounded by a membrane and contain chlorophyll, DNA, ribosomes, and a large quantity of enzymes. **Chlorophyll** is a pigment capable of absorbing light and storing it in the form of chemical compounds.

The chloroplasts are located primarily in the leaves. If we cut a cross section of a leaf, we can see the cells with a magnifying glass; together, these cells make up the tissues. The cells contain a large number of chloroplasts. That is what makes leaves the main producers of oxygen.

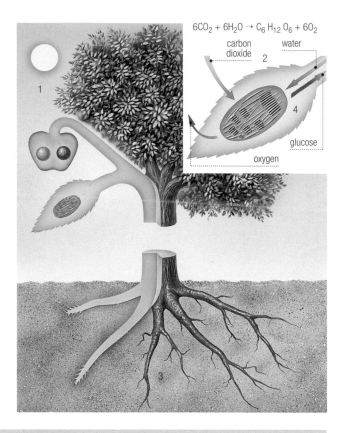

$$6CO_2 + 6H_2O \rightarrow C_6H_{12}O_6 + 6O_2$$

carbon dioxide 2 water

4

glucose

oxygen

1

In addition to their job of purifying the air, plants provide humans with food (both wild and cultivated), in addition to fibers, wood, and other very useful items.

Photosynthesis takes place inside the leaves; this is a chemical reaction that combines carbon dioxide and water to produce glucose.

1. The plant synthesizes nutritive substances through photosynthesis.
2. This process is carried out in the leaves.
3. It uses water absorbed through the roots.
4. The chlorophylls, green pigments located in the chloroplasts, capture the energy from the sun.

PHOTOSYNTHESIS

Photosynthesis involves two principal phases: the light phase and the dark phase.

In the middle of the seventeenth century, a Dutch doctor, van Helmont, conducted a curious experiment: He planted a willow tree in a pot, weighed the quantity of earth precisely, and isolated it from contact with the outside, leaving holes only for water drainage. At the end of five years, he repeated his measurements; the dirt weighed nearly the same, but the willow had gained more than 150 pounds (70 kilos). Van Helmont concluded that the weight gain was due solely to water. At the end of the following century, it was determined that an increase in plant mass is due to gases from the air and the presence of light. The tree had manufactured plant material with the help of these gases, water, and light. The process was named photosynthesis (photo = light; synthesis = production). Today, we know that it is not just one reaction but a series of many.

Although all plants need light, they do not all need the same amount.

Photosynthesis takes place in two main stages: a set of reactions that occur in the dark makes up the dark phase (which requires no light), and another set that occur in the presence of light makes up the **luminous phase**. During the latter, thanks to the solar energy, water breaks down into hydrogen and oxygen and forms **ATP** molecules, which store energy. Then the energy molecules are used in the dark phase to attach the **carbon dioxide** (CO_2) to the water and form glucose.

CHLOROPHYLL

Chlorophyll is a pigment that has the ability to absorb solar radiation. In fact, it is a combination of two similar pigments, **chlorophyll a** and **chlorophyll b**, which are distinguished from one another by the color of light—the range of wavelength—that they absorb. Chlorophyll is found inside the **chloroplasts**; it is responsible for the light phase of **photosynthesis**: The energy it absorbs serves to break down water and form **ATP** molecules. Chloroplasts contain other pigments of different colors, besides chlorophyll, that absorb different wavelengths of light. Prominent among them are the carotenoids, which go from red to yellow.

SPECTRA OF ABSORPTION OF CHLOROPHYLL A AND B

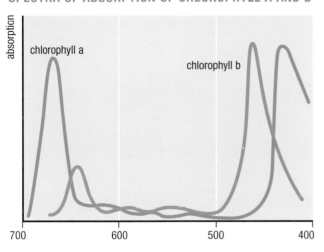

chlorophyll a

chlorophyll b

absorption

700   600   500   400

NITROGEN FIXATION

During photosynthesis, plants manufacture organic material, but the overall manufacturing process also includes another very important reaction: **nitrogen** fixation; in other words, producing a compound that contains nitrogen. This is a component in the **proteins** that all living beings, including plants, need. Certain bacteria, especially the genus *Rhizobium*, fix atmospheric nitrogen in the form of **nitrates**; this process consumes lots of energy on the part of the bacteria that live in a symbiotic relationship on the roots of certain plants—mainly leguminous ones such as lentils and alfalfa. Plants that do not have these bacteria have to absorb the nitrogen in the form of nitrates that exist in the soil, and they die if they cannot get them.

leguminous plant

The nitrogen-fixing bacteria (*Rhizobium*) live by forming nodules on the roots of leguminous plants.

bacteria nodules

Chlorophyll is green because it reflects the green light that it does not use.

Chlorophyll is one of the green substances that gives color to the young aerial parts of most plants.

ANIMAL CHEMISTRY

Like all living beings, animals are made up of chemical substances organized into chemical structures (cells, tissues, organs, etc.) inside which chemical reactions take place. These reactions are the ones that keep the creatures alive, since they transform the foods into material for their bodies, produce heat, make movement possible, and so forth. Other substances are used in controlling the whole process.

METABOLISM

In order to live, animals need **energy**. They get it from the organic material that makes up their food. In some cases, this is plant matter (as with **herbivores** such as cows), and in others, it is animal matter in the form of other prey animals (as with **carnivores** such as wolves). Inside their bodies, the food is broken down into simple compounds that nourish the cells. All of this takes place through **chemical reactions**. All of these reactions taken together are referred to as **metabolism**.

All the substances that are produced in the course of metabolic reactions are called **metabolites**.

DIAGRAM OF THE MAIN REACTIONS THAT TAKE PLACE INSIDE AN ANIMAL

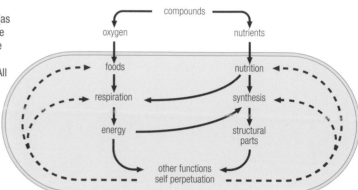

THE BODY: A CHEMICAL FACTORY

Both respiration and cellular nutrition are chemical reactions, since the **cells** act like small factories that extract energy from foods and use it for growth and reproduction, giving rise to new cells that shape the animal's body. The cells form **tissues** and **organs**, and each one of these highly organized structures performs a certain function (absorbing oxygen from the air, transforming nutrients, producing reproductive cells, etc.). Still, in order for the set of organs that make up the body to function in coordination with one another, a control system is needed (for sending orders, transporting messages, and so forth). This takes place with the aid of two systems: one is the nervous system, and the other is of a chemical nature dictated by the substances called **hormones**.

The body of a cow is made up of millions of cells that work in harmony with one another because of the hormonal control system.

INSECT HORMONES

From the time an insect is created in the egg to the appearance of the adult phase, it experiences a series of transformations that constitute the phenomenon known as **metamorphosis**. In addition, insects have an exterior skeleton similar to a shell; as a result, in order to grow, they have to change their skeleton and make a new, larger one. The changing of this external covering is known as **shedding** or **molting**. Both processes occur with amazing precision because of a series of hormones that work together to coordinate them. In each stage of the metamorphosis, the internal organs (such as the encephalon or brain) produce hormones that trigger the next step. The same thing occurs when insects shed their exoskeletons.

One biological means of combating infestations involves supplying insects with hormones that interrupt their life cycle.

The thyroid gland produces tyrosine; in amphibians, this substance controls metamorphosis; in humans, a lack of it causes goiters.

THE TRANSFORMATION FROM CATERPILLAR TO BUTTERFLY

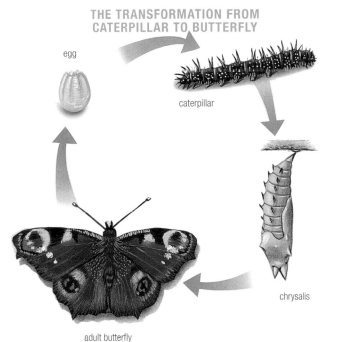

egg

caterpillar

chrysalis

adult butterfly

The chemical regulatory system that works by means of hormones is called the **endocrine system**.

MAMMALIAN ENDOCRINOLOGY

ORGANS THAT SPECIALIZE IN HORMONE PRODUCTION

anterior lobe

posterior lobe

pituitary gland

parathyroid

thyroid gland

cortex

medulla

adrenal gland

follicle

corpus luteum

ovary (females only)

endocrine cells

generative cells of spermatozoids

testicle (males only)

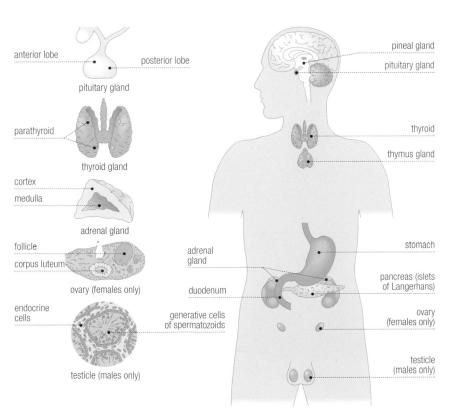

pineal gland

pituitary gland

thyroid

thymus gland

adrenal gland

stomach

duodenum

pancreas (islets of Langerhans)

ovary (females only)

testicle (males only)

Mammals, such as human beings, are very complex animals with many organs and organ systems. They are subject to double controls through the nervous system on the one hand, and a highly developed hormone system on the other. There are several organs that specialize in the production of hormones that are crucial to the individual's development. Among them are the **hypophysis** or pituitary gland; this gland is very important, since it coordinates the workings of nearly all the other glands. The **thyroid** controls organic metabolism by regulating the amount of heat that the body produces. The **pancreas** produces **insulin**, which regulates the amount of glucose in the blood. The **testicles** produce **androgens** that determine the male secondary sexual characteristics. The **ovaries** produce such hormones as **progesterone**, which prepares the uterus for pregnancy. As we can see, all animal activities are controlled by hormones.

The pituitary gland produces many hormones, including the human growth hormone (HGH) and the follicle-stimulating hormone (FSH), which produces estrogen.

THE HISTORY OF LIFE ON OUR PLANET

From the time of the emergence of life on our planet, some 3.5 to 3.6 billion years—or perhaps even more—have passed. The first organism was a **molecule** of organic matter that developed the ability to grow, relate to other molecules, and reproduce. Evolution has been going on since that moment, and it has resulted in increasingly complex organisms.

The first ones were **single-celled** marine organisms. Later on, they developed into **multicelled** marine organisms; eventually conditions outside the water improved to the point where they could leave the water. That is how land **plants** and **animals** began to make their appearance; they have developed all kinds of shapes and ways of surviving.

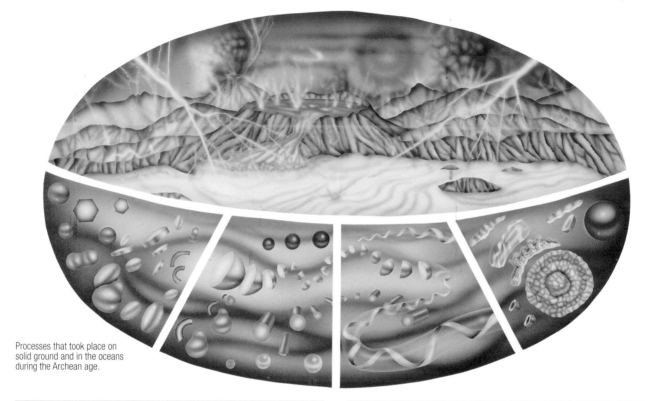

Processes that took place on solid ground and in the oceans during the Archean age.

THE ARCHEAN AGE

This is the oldest epoch, since it began some 3.6 billion years ago. It seems that there was great volcanic activity, huge storms, and very severe erosion of exposed land. This is the time when the first organisms appeared.

THE PROTEROZOIC AGE

This age began around 1.6 billion years ago. The formation of the most ancient glaciers took place during this time. Erosion was very severe, and volcanic activity resumed toward the end. The sea became populated with worms, jellyfish, and sponges, along with various types of aquatic plants.

Scene from the Permian period.

Scene from the Paleozoic era.

THE PALEOZOIC ERA

This era began some 600 million years ago. At the beginning it was warm; afterward, however, the dryness increased, and new glaciers were formed. This era is divided into six periods with distinguishable characteristics. Marine fauna diversified a lot; chief among them were the **trilobites**, and among the vertebrates, the **cartilaginous fish**. Insects began their conquest of terra firma; in the middle of the era, the **amphibians** made their appearance, followed by **reptiles**, and, toward the end, the first **dinosaurs**. Land plants diversified greatly and very leafy **forests** sprang up; they are the source of present carbon deposits.

Era	Period	Duration (millons of years)	Start (millons of years)
Archean		2,000	3,600
Protcrozoic		1,000	1,600
Paleozoic	Cambrian	100	600
	Ordovician	75	500
	Silurian	20	425
	Devonian	60	405
	Carboniferous	65	345
	Permian	50	280
Mesozoic	Triassic	50	230
	Jurassic	45	180
	Cretaceous	70	135
Cenozoic	Tertiary	64	65
	Quarternary	1	1

The brontosaurus was a dinosaur of the Paleozoic era. It was some 70 feet (22 meters) long, and it is estimated that it weighed around 30 tons.

THE MESOZOIC ERA

This era began around 230 million years ago. In many instances, the oceans covered the continents, and, in the middle of this age, there was great activity in forming mountains. Many of the oldest animals disappeared, and in the oceans the dominant creatures were **cephalopods**. On solid ground there was a great diversification among the insects. This is the era in which the first **birds** and **mammals** appeared, and it was the highest point in the existence of the dinosaurs, which became extinct at the end of the Mesozoic era. In the plant world, the **phanerogams** began their dominance.

The Cenozoic era began around 65 million years ago, and it included several periods of **glaciation**. The continents were taking on their present shape, **monocotylodonous** plants made their appearance, and **placental mammals** began their dominance.

The mammals, such as this platybelodon, a precursor of the elephant, made their appearance in the Mesozoic era.

FOSSILS: THE HISTORY OF LIFE

Today, we know that present species have developed from preexisting ones through the process of evolution, but this knowledge became available only through the study of the remains of those organisms. They have provided the proof that evolution has been taking place on our planet since life first appeared. Paleontology is the science that studies these remains, the fossils.

PALEONTOLOGY

Paleontology gained a lot of ground in the nineteenth century when fossil remains of huge animals kindled interest among scientists, even though fossils had been known since antiquity. Paleontology attempts to discover what the plants and animals of the past were like based on the remains that have survived to our time. To do this, they compare the anatomical characteristics of the various groups and use geology to identify different ages. Two of the important branches of paleontology are **paleobiology**, which studies fossils from a biological viewpoint, and **paleoecology**, which focuses on the ecological conditions of the era in which the fossilized organisms lived.

Trilobites are one type of fossil that is fairly easy to find. They were a kind of arthropod from about one to three inches (3 to 10 cm) long; they inhabited the shores and the shallow areas of the ocean during the Paleozoic era.

One very important thesis of paleontology (actualism) holds that the fossilized species were governed by the same biological laws as present-day species.

Petrified trees in the Patagonia region of Argentina.

Fossils help us understand what the plants and animals were like millions of years ago.

FOSSILIZATION

When an organism dies, its remains generally decompose and disappear. However, there are circumstances in which things turn out differently. The remains are protected from rotting and end up turning into a **fossil**. The process is known as fossilization; it results in the gradual replacement of organic material by inorganic material through **chemical reactions**. That way, we find the fossil in the form of a rock in which all the organs or the parts of a plant or animal have been transformed into a mineral but without losing their shape.

The presence of fossils of tropical species in cold regions is evidence of climatic changes and of continental drift.

EVOLUTION AND GENETICS

TYPES OF FOSSILS

Normally fossils preserve only the hard parts of an animal, in other words, its **skeleton**, its **shell**, and so forth. In other cases, what we find is the **track** or the **mold** of its presence. What happens is that the organism ended up in soft material (such as mud) that later was transformed into rock and preserved a hollow in the shape of the body. The tracks that animals left as they walked across a soft surface also fall into this category. The substances of organic origin that are found in sedimentary terrain and reveal what types of organisms lived there are called **chemical fossils**.

Reconstructed fossil of a dinosaur.

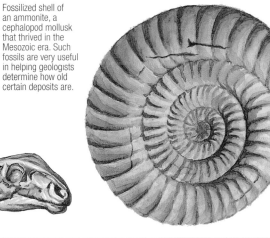

Fossilized shell of an ammonite, a cephalopod mollusk that thrived in the Mesozoic era. Such fossils are very useful in helping geologists determine how old certain deposits are.

LIVING FOSSILS

This is the term applied to certain animals and plants that are living today, because it was thought that they had become extinct or because they belong to a very ancient genus of which they are the only surviving representatives. One of the best-known examples is the coelacanth, a fish discovered at the beginning of the twentieth century in the Indian Ocean; people had thought that the fish disappeared more than a hundred million years ago. It has very primitive characteristics, and it has been very helpful in proving theories about what the members of this group must have been like during the **Mesozoic Era**.

The coelacanth is the only representative of the coelacanthiforms, a group of fishes that disappeared in the Cretaceous period.

Scientists captured the first living coelacanths during the 1930s.

Ice is a good preservative; it has made it possible to find entire animals from thousands of years ago, such as woolly mammoths.

The meat from woolly mammoths frozen in glacial ice could still be eaten today.

EVOLUTION AND ITS MECHANISMS

A process known as evolution is responsible for the tremendous variety of organisms that have evolved from a single tiny cell and that now inhabit the planet. Evolution takes place through small changes that appear in isolated individuals and that eventually lead to the appearance of new species. The mechanisms by which evolution operates are extremely varied but relatively simple.

Islands such as the Galapagos are a great place for the development of new species. The picture shows blue-footed boobies that are characteristic of the Galapagos.

SPECIATION

Speciation is the **formation** of a species—in other words, the appearance of a set of organisms that have their own characteristics and are capable of reproducing. One of the most common means by which new species appear is through **isolation**. One example is the **Darwin's finch**. This finch originally inhabited the American continent and migrated to the Galapagos Islands. Evolution changes organisms gradually, but since the finches on the islands were subjected to different conditions than the mainland finches, their evolution followed a different track. The result was a new species of finches.

 In 1859, Charles Darwin published his book *On the Origin of the Species*, in which he expounded his theory of evolution.

PROOF OF EVOLUTION

There are numerous proofs of the validity of evolution. The science of **paleontology** has contributed many data through the study of fossil remains. In many species, there are fossil forms from the various stages of their evolution that show the gradual changes that have led to the present. **Comparative anatomy** is extremely helpful; it consists of comparing anatomical parts of different species to establish their similarities and differences and determine how related they are.

Embryonic development also demonstrates evolution, for every animal in the embryonic state passes through the stages that its species has experienced through the ages; for example, the human embryo has a tail, as the other primates do, but it subsequently disappears. Modern technology has contributed new proofs through the study of **chromosomes** and the **biochemistry** of organisms.

One example of change through mutation may involve a larger beak in a bird, allowing it to feed on insects that live buried in the soil. If the change proves favorable, it will be transmitted to descendants; but, if it is not favorable (for example, because the bird lives in a place where the ground is rocky), it will disappear and the individual that exhibits it will die without transmitting the characteristic to its descendants.

VERTEBRATE EVOLUTION

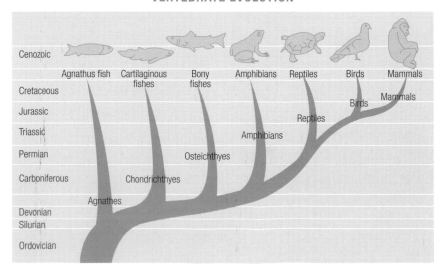

Cenozoic — Agnathus fish / Cartilaginous fishes / Bony fishes / Amphibians / Reptiles / Birds / Mammals
Cretaceous — Birds / Mammals
Jurassic — Reptiles
Triassic — Amphibians
Permian — Osteichthyes
Carboniferous — Chondrichthyes
Devonian — Agnathes
Silurian
Ordovician

EVOLUTIONARY MECHANISMS

Evolutionary mechanisms are the factors by which an individual or a species acquires a new characteristic that differentiates it from its predecessors. But these changes must be transmittable to offspring, otherwise they disappear with the death of the individual. The changes need to be in the **genetic material** and that happens through **mutations**. Some factor (such as temperature or radiation) causes a change in the genes that is subsequently passed on to descendants. Then **natural selection** comes into play; this allows the survival of the individuals that are best suited to given conditions.

PLANT EVOLUTION

Aquatic plants

Mosses and liverworts

Lycopodia

Ferns

Flowering plants

Cycadaceae

Conifers

Gingko

Equiseta

Fungi

Angiosperms

Phylocenes

Gymnosperms

Lycopsida

Sphenopsida

Pre-ferns

Tracheophytes

Bryophytes

Multicelled aquatic plants

First cells

NATURAL SELECTION

In his theory of evolution, **Darwin** spoke of natural selection as the survival of the fittest. This means that nature guides the process of evolution, since it allows to survive only those individuals (and their species) that adapt successfully to environmental conditions. However, the Earth has many different environments, and that is why there are so many species. Natural selection acts equally on all individuals in a population. Some of them will exhibit small differences (resulting from **mutations**), and, if they are advantageous, those individuals will reproduce more successfully. As a result, the number of individuals with these characteristics increases, and the species undergoes evolution. A change in environmental conditions (such as an ice age) favors the individuals that have characteristics suited to the new situation, and their population will achieve dominance.

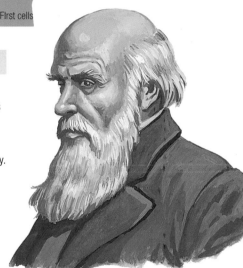

In 1831, shortly after graduation, Darwin set sail on the *Beagle* and traveled for five years (continually seasick, according to his diary) through South America and the Pacific Islands making scientific observations; these data formed the basis of his famous theory of the evolution of species.

Extreme conditions high in the mountains demand the development of plants and animals that are capable of adapting.

THE BASIS OF HEREDITY

The elongated form of a worm, the red and white flowers of a rosebush, an elephant's trunk, and the color of our eyes and hair are characteristics that are echoed in different individuals in every species. They are characteristics that are inherited, and every one of them is "made" in accordance with instructions that the genes provide to the cells whose job it is to produce tissues and organs.

GENETICS

This science, which is so important and appreciated in today's media, is devoted to the study of how characteristics are transmitted from one generation to another. In the middle of the nineteenth century, Mendel discovered the laws that govern inheritance, but genetics did not really get started until the twentieth century, when the physical elements that make that possible—genetic material—were discovered.

The **genotype** is the genetic makeup of an individual based on one or more characteristics—for example, the genes that determine eye color.

Mendel founded the science of genetics by discovering the laws of heredity.

GENES

A **gene** is defined as the hereditary unit of material that is transmitted from one generation to the next and that is capable of undergoing **mutations**, that can be **recombined** with other genes, and that can determine the nature of the organism.

The **phenotype** is the external manifestation of the genotype—for example, the color determined by the eye color genes.

TYPICAL CHROMOSOME STRUCTURE

chromatid

arm

telomere

centromere

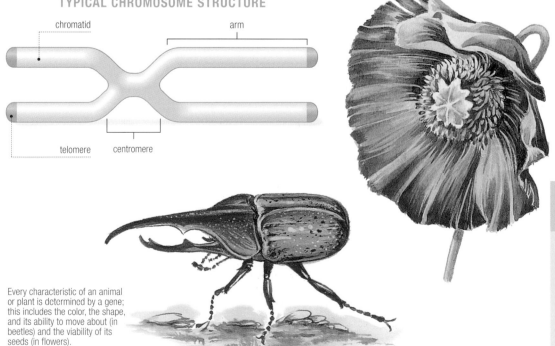

Every characteristic of an animal or plant is determined by a gene; this includes the color, the shape, and its ability to move about (in beetles) and the viability of its seeds (in flowers).

The word *heredity* comes fr the Latin *haerentia*, which means *things that are linke together* or *belonging*; the word *genetics* comes from Greek *genesis*, which mean *origin* or *creation*.

CHROMOSOMES

A chromosome is a structure inside the **nucleus of a cell**; it can easily be colored and can be observed while it is dividing. A chromosome is made up primarily of deoxyribonucleic acid, or **DNA**, and proteins. It is fairly long in structure and contains a pair of arms joined at one point, either in the middle or at one end. The chromosomes carry the genetic material (the **genes**), and their number varies; however, it is always the same for each species.

THE NUMBER OF CHROMOSOMES IN VARIOUS SPECIES

Species	Number of chromosomes
Ascarids	3
Fruit fly	8
Mushroom	8
Corn	20
Snake	36
Humans	46
Rosebush	56
Carp	104
Water lily	112
Ophioglosum	520

The DNA has the capability of replicating itself. Each of its two filaments creates a new one.

DNA

Deoxyribonucleic acid, known as **DNA**, is an essential component in chromosomes, and it is the substance that makes up the **genes**. It is a polymer—in other words, a giant molecule made up of a great number of smaller units, the **nucleotides**. However, it contains just four nucleotides, but this is all it needs for combining with itself. This macromolecule consists of two filaments that are coiled around one another to form a sort of spiral staircase: This structure is known as a **double helix**.

DIAGRAM OF PROTEIN SYNTHESIS

DNA

polypeptide chain in formation

amino acids

In Tri Sar → Ala

RNA-m

from the nucleus to the cytoplasm

ribosome

Double helix structure of DNA.

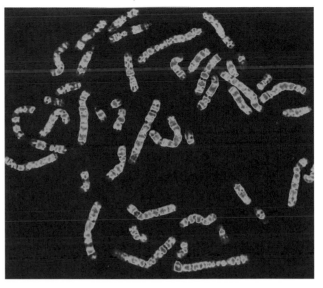

Human chromosomes seen under an electron microscope.

GENETIC CODE: A UNIVERSAL LANGUAGE

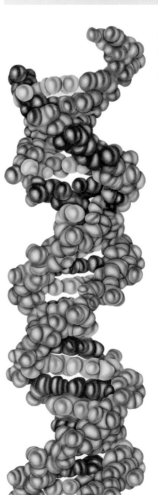

Genetic information is contained in the **DNA**, and, in order for **protein synthesis** to take place, the DNA produces a messenger whose job it is to deliver the orders. The messenger is the **RNA-m** (messenger RNA); it is an elongated macromolecule that carries copies of the information from the DNA. When it arrives at a ribosome, it attaches to it and the protein synthesis begins.

The orders are transmitted through a combination of bases. DNA has four different ones (A, G, C, and U), which can combine a total of 64 different ways. Since only 20 **amino acids** are needed to produce proteins, that number is ample. The various combinations using three bases are what is known as the **genetic code**.

In 1953, James Watson and Francis Crick discovered the double helix structure of DNA.

THE LAWS OF HEREDITY

The characteristics of all individuals are determined by their genes. In organisms that reproduce sexually, each individual is the product of the joining of two sexual cells, each one of which brings its own genes. The result is a combination that displays, to a greater or lesser degree, the characteristics of the progenitors. The explanation for this is provided by the discoveries of Mendel.

THE TRANSMISSION OF CHARACTERISTICS

Every characteristic is controlled by a **gene**, and most higher organisms have two genes for each characteristic; in other words, they are **diploid**. This is represented by a pair of letters such as AA. This is what is known as a **pure characteristic** (for example, the color white); another could be BB (for instance, the color black). But there can also be **hybrid characteristics**, such as gray, which is thus represented by AB. However, not all genes are equally influential when it comes to producing their characteristic. As a result, if the AB hybrid is not gray, but rather white, we say that white is **dominant** and black is **recessive**. Since these genes affect a single characteristic (color), they are represented by the same letter, and thus white is AA, black is aa, and the hybrid is Aa.

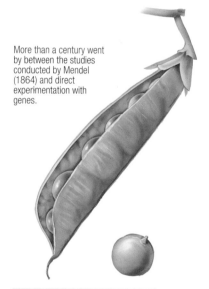

More than a century went by between the studies conducted by Mendel (1864) and direct experimentation with genes.

In a single flock of sheep, there are white individuals and black ones.

MENDEL'S EXPERIMENTS

In order to discover his laws, **Mendel** used **pure strains** of peas that differed in only one respect (green or gray, with a smooth or rough surface). That way, the results were clear and it was possible to deduce how the genes combined. But they are also applicable to complex cases in which several characteristics are mixed together.

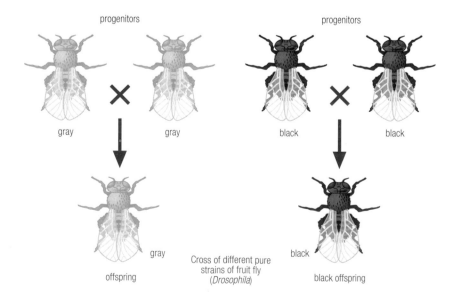

progenitors

gray   gray

gray

offspring

progenitors

black   black

black

Cross of different pure strains of fruit fly (*Drosophila*)

black offspring

MENDEL'S FIRST LAW

When two pure strains that differ in only one characteristic are combined, all their descendants are the same and their phenotype coincides with that of one of the progenitors.

For example, in crossing gray flies and black flies, all the descendants are gray. This is due to the fact that the gray trait for body color is dominant over the black trait.

MENDEL'S SECOND LAW

When two hybrid individuals are crossed (generation 2, which is the product of generation 1 involving purebred individuals), the descendants exhibit the phenotypes of the first generation in a fixed proportion.

For example, by crossing hybrid gray flies (resulting from crossing gray flies with black flies), a fourth of the descendants are black and three quarters of them are gray.

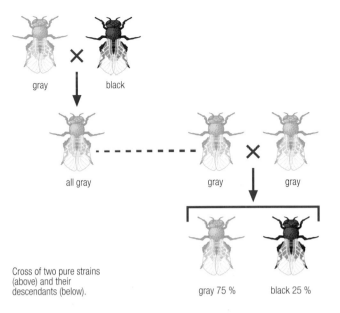

gray   black

all gray   gray   gray

Cross of two pure strains (above) and their descendants (below).

gray 75 %   black 25 %

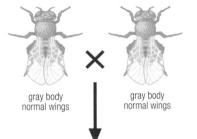

Cross of two individuals of the second generation, which transmit traits independently.

gray body   gray body
normal wings   normal wings

gray body   gray body   black body   black body
normal wings   vestigial wings   normal wings   vestigial wings

MENDEL'S THIRD LAW

Each of the two alternatives of a characteristic is transmitted independently of the other characteristics of a hybrid. In other words, the inheritance of one characteristic is generally not affected by the inheritance of a different characteristic.

One exception may involve the inheritance of a harmful trait (such as a fatal illness), which may produce an unexpected result from the cross. An example of the third law is a cross between hybrid flies with normal wings and a gray body. The descendants will exhibit four different phenotypes in proportions that are also different.

Many illnesses that were unexplainable in the past can now be addressed using discoveries from genetics.

Because of heredity, successive generations exhibit not only some common physical traits but also traits involving character, habits, and values.

BIOTECHNOLOGY AND GENETIC ENGINEERING

One of the most spectacular applications of genetics is genetic engineering, which offers tremendous possibilities, even though it may also cause unknown problems. It was made possible through the great advances of the second half of the twentieth century, and, in conjunction with biotechnology, it is now an important industry.

BIOTECHNOLOGY

This concept is widely used nowadays, and it is sometimes abused. Biotechnology involves taking advantage of the metabolic abilities of living beings to produce useful products. Nowadays, this often is a sophisticated technique, but, in fact, it is very old, since many foods currently produced such as bread, wine and beer, yogurt, and cheese are little more than the results of using certain organisms to produce a product for human consumption.

Cheeses, which are now made on an industrial scale, are one product of biotechnology.

GENETIC ENGINEERING

In theory, this practice is very simple, but it is very difficult to put into practice because it involves working with elements of very minuscule proportions. It consists of manipulating an organism's **genetic material** in order to change its characteristics. In some cases, **defective genes** are removed and replaced by new ones to give the organism some new characteristic. For that purpose, its **chromosomes** have to be extracted, and sections of DNA that correspond to a certain gene have to be cut out. This segment can be replaced by another in cases where a gene is defective.

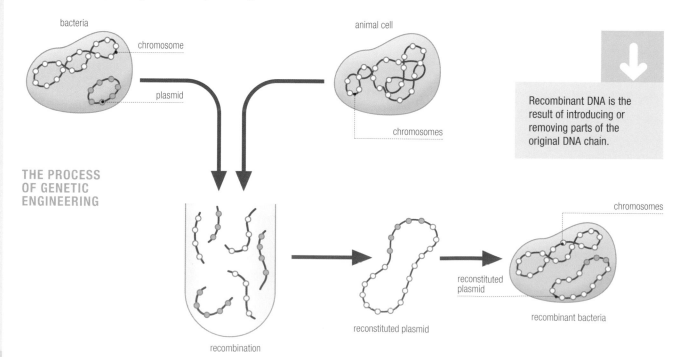

Recombinant DNA is the result of introducing or removing parts of the original DNA chain.

THE PROCESS OF GENETIC ENGINEERING

bacteria
chromosome
plasmid

animal cell
chromosomes

recombination

reconstituted plasmid

reconstituted plasmid

chromosomes
recombinant bacteria

THE POTENTIAL OF GENETIC ENGINEERING

It is possible to eliminate genes that cause illnesses by manipulating the genetic information in an individual's cells. This is a very encouraging field for **medicine**, and it may make it possible to cure many fatal illnesses that now plague humanity.

In **agriculture** and **animal husbandry**, it is possible to introduce genes that favor some special characteristic that has commercial value. This is already being done with many plants, such as corn. Genetically modified (transgenic) strains of corn have been created that can resist parasites. There are varieties of tomatoes that grow larger and look good for a longer time, which is of great commercial interest.

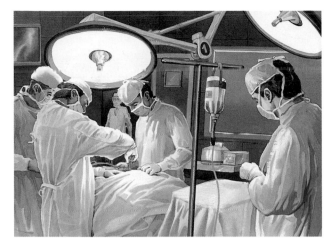

Genetic engineering opens up possibilities of obtaining tissues and organs for transplants without triggering rejection by the recipient, as it often happens.

Clones are individuals that carry precisely the same genetic information as their progenitors; as a result, they are identical to them. In asexual reproduction, all descendants are clones, in contrast to sexual reproduction, in which different characteristics are combined.

At the end of 2001, an American laboratory succeeded in cloning human embryos for the purpose of obtaining tissues for medical purposes.

One of the most famous attempts at animal cloning involved the sheep Dolly.

ARE TRANSGENIC ORGANISMS SAFE?

The main objection offered by many people to genetically modified organisms is their safety. In principle, transgenic foods are safe for our health because they possess the necessary nutrients and have the same characteristics. Still, the introduction into nature of plants and animals that have been modified in such ways involves a risk in not knowing how they will fit into the ecosystems (as happened with DDT before it was banned). In addition, transgenics that have accidentally been freed (and experience shows that this is not an impossibility) could turn into carriers of new diseases for which we are not prepared.

Corn is one crop with which a fairly large number of transgenic varieties have been developed.

CELLS

Plants and animals come in all sizes, but all of them, from tiny creatures to whales and sequoias that are hundreds of years old, are made up by a variable amount of the same units, cells. A cell is defined as the smallest unit of life, and, in fact, there are organisms, such as protozoa, that are made up of just one cell.

 Cells that have a true nucleus are called eukaryotic; cells that do not have a definite nucleus are called prokaryotic.

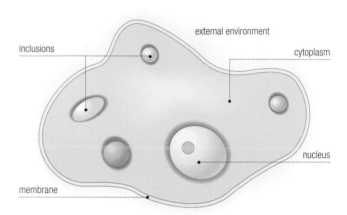

external environment

inclusions

cytoplasm

nucleus

membrane

THE CELL, THE UNIT OF LIFE

The cell is considered to be the basic structural unit of all living beings. It is the minimum amount of organized living matter that is capable of growing and multiplying. Essentially, a cell consists of a dense material known as **cytoplasm** surrounded by a **membrane** that separates it from the outside, and of a series of structures inside it, including the **nucleus**, where the genetic material is located. However, there are some very primitive organisms that do not even have a differentiated nucleus, and their cell thus consists of the membrane that surrounds the cytoplasm and all its inclusions.

Ribosomes are very tiny spheres (visible only through an electron microscope) that play a role in protein synthesis.

THE COMPONENTS OF A CELL

One of the most important components of a cell is the **membrane**, which sometimes is provided with a rigid wall. The membrane has a double function: It has to insulate the interior from the exterior, but, at the same time, it has to allow an exchange of materials between the cell and the environment that surrounds it. Therefore, it is semipermeable; this means that it allows certain substances and materials to pass through but not others. In addition to the **nucleus**, the **cytoplasm** also contains **mitochondria**, **chloroplasts**, **ribosomes**, **endoplasmic reticulum**, **Golgi apparatus**, **lysosomes**, and **vacuoli**.

Mitochondria are spheres or small rods that contain the enzymes that are responsible for the oxidation of foods.

The **endoplasmic reticulum** is a system of membranes that sometimes carries attached ribosomes.

Cross section of a Golgi apparatus (above) and diagram showing the structure of a cell membrane (below).

A **Golgi apparatus** is a stack of flat sacs surrounded by membranes; it is used in the synthesis of polysaccharides.

Lysosomes are spherical structures surrounded by a membrane and that are filled with proteins. They contribute to the incorporation of foods into the cell and to its disintegration when it dies.

Vacuoli are bubbles filled with various materials that are either useful or discarded.

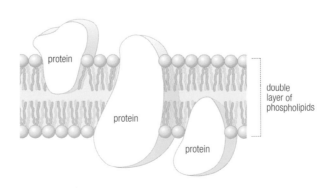

protein

protein

protein

double layer of phospholipids

CILIA AND FLAGELLA

These structures are extensions of the membranes; inside, they contain a series of filaments (known as microtubules) arranged concentrically around one in the center. They can be short ones known as **cilia** or long ones referred to as **flagella**. They also vary in number; there are many cilia, but few flagella, often only one. The function of both structures is to generate water currents and help the cell move around.

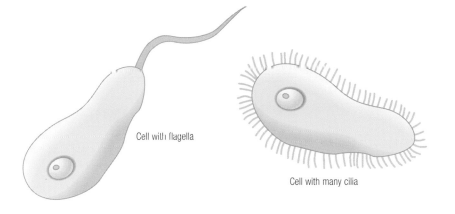

Cell with flagella

Cell with many cilia

GENERAL DIAGRAM OF AN ANIMAL CELL

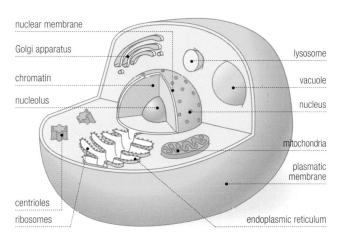

- nuclear membrane
- Golgi apparatus
- chromatin
- nucleolus
- centrioles
- ribosomes
- lysosome
- vacuole
- nucleus
- mitochondria
- plasmatic membrane
- endoplasmic reticulum

ESQUEMA GENERAL DE UNA CÉLULA VEGETAL

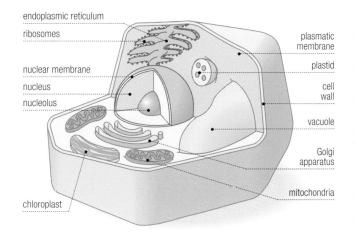

- endoplasmic reticulum
- ribosomes
- nuclear membrane
- nucleus
- nucleolus
- chloroplast
- plasmatic membrane
- plastid
- cell wall
- vacuole
- Golgi apparatus
- mitochondria

 Chloroplasts contain the chlorophyll that makes photosynthesis possible.

TYPES OF CELLS

All cells have the same basic structure, but there are some components that appear in some cells and are absent from others. This allows making a primary distinction between **plant cells** and **animal cells**. The former have a more or less rigid cellulose **wall** that surrounds the membrane, chloroplasts, and other plastids and large vacuoles; the latter lack the cell wall, the chloroplasts, and the plastids, but they have lysosomes, and their vacuoles are small. In addition, the cells are differentiated by function when they make up tissues, whether or not they have cilia and flagella, and so forth.

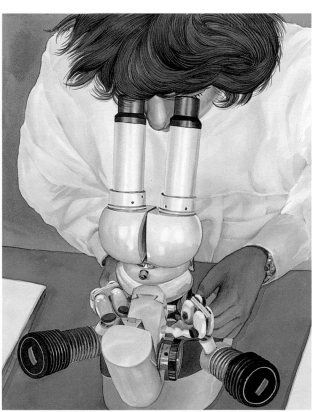

The microscope has proven indispensable in studying items as tiny as cells and microorganisms. It was invented by the Dutch scientist Z. Jansen at the end of the sixteenth century; today, we have extremely powerful electron microscopes that help scientists understand the secrets of life.

TISSUES

The simplest organisms consist of a single cell that performs all the necessary functions. The rest of the organisms are made up of a variable number of cells, and, in most cases, except in the least highly evolved species, all these cells are specialists in certain functions and thereby make up what we refer to as tissues.

THE SPECIALIZATION OF CELLS

When a cell specializes by performing a certain function, it changes shape, it loses some of its components, and it takes on, or reinforces, others so it can work more effectively. This is a strategy used by living things to increase their efficiency: **division of labor**. The result is groups of cells equally specialized in some activity; these cells are known as **tissues**. Every organism produces a particular type of tissue that is adapted to the type of life it leads. Thus, many trees produce protective tissues such as cork; aquatic plants produce tissues that act like flotation devices; and animals produce tissues that are capable of contracting so that they can move about.

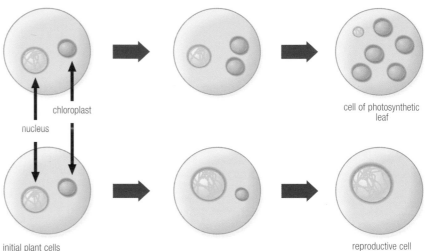

chloroplast

nucleus

cell of photosynthetic leaf

initial plant cells

reproductive cell

The leaf cells devoted to photosynthesis come from the same type of cell as the ones used for reproduction.

The process of specializing in a certain function is called **cellular differentiation**.

PLANT TISSUES

Plants produce various types of tissues; sometimes they are made up of a single type of cells and, at others, by several. The main types are **embryonic** tissues, **protective** tissues, **support** and **nutritional** tissues, and **conductive** tissues.

DIFFERENT TYPES OF PLANT TISSUES

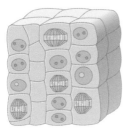

meristematic

Embryonic or **meristematic** tissues make up meristems and consist of nondifferentiated cells located in the plant's growth areas (vegetative apexes).

The tissues used for **support** and **nourishment** make up a major part of the plant's mass; they give it strength and allow it to store up reserve substances. The main ones are **parenchyma**, **cholenchyma**, and **sclerenchyma**.

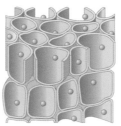

protective

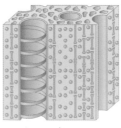

parenchyma

xylem

Protective tissues help ward off injury; they are made up of flattened cells, often with a reinforced wall. They cover the leaves, the stems, and the roots.

Conductive tissues are used for transporting liquids and materials throughout the plant. These are the tubes that go through the roots, the stem, and the leaves, and that transport sap. The main ones are the **phloem** and the **xylem**.

ANIMAL TISSUES

Animals possess a higher degree of specialization than plants, and their tissues are also more diverse. For example, in a higher vertebrate, it is possible to distinguish more than a hundred different types of specialized cells that make up tissues. However, they all can be grouped into four large categories: **epithelial** tissues, **conjunctive** tissues, **muscle** tissues, and **nervous** tissues.

DIFFERENT ANIMAL TISSUES

epithelial

scaly

columnar

cuboidal

The **epithelial** tissues consist of flattened cells that are tightly clustered to form a compact surface. They perform a great many functions: They cover the body to protect it from outside agents, and they also cover the internal cavities to perform the exchange of materials (in the intestine) and gases (in the lungs and bronchial tubes).

muscular

nervous

The **conjunctive** tissues are made up of differentiated cells surrounded by a great quantity of extracellular material. These include the **cartilaginous** and **bony** tissues, the **tendons**, and the **ligaments**.

cartilage

ligament

The **sexual** cells originate in the germinal tissue, a specialized epithelial tissue.

bony

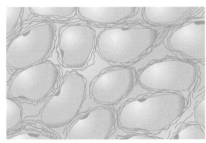
adipose

conjunctive

The **nervous** tissues are made up of cells that specialize in transmitting electrical impulses; these cells are known as **neurons**, and they are provided with many extensions that connect them to one another. The spinal medulla and the brain in vertebrates are made up of many neurons.

The **nerves** are extensions of the neurons.

The **muscle** tissues are made up of elongated cells that have the capacity to contract. There are two main types: **smooth muscle tissue** (for involuntary contractions) and **striated muscle tissue** (for voluntary contractions).

The **cardiac** tissue is a specialized muscular tissue.

The **adipose** tissue is a specialized conjunctive tissue whose cells are filled with fat.

ORGANS

The vast majority of multicelled organisms have more or less specialized tissues. In addition, as they evolve and their bodies become more complex, these tissues group together in larger units known as organs, which perform more complex tasks than those carried out by a single tissue. In the most highly evolved organisms, the organs are, in turn, assembled into units known as systems.

PLANT ORGANS

In the lower plants, the body is known as a **thallus**; it has no organs, but, in some cases, such as the mosses, it takes on a form similar to that of the higher plants. In the latter, the body is divided into three large, main organs: the **root**, the **stem**, and the **leaves**. The main job of the roots is to absorb water and nutrients, but they also anchor the plant in the substrate. The stem contains the conductors of the sap, and it holds up the plant and gives it shape. The leaves are the part that generally carry out **photosynthesis**.

Sometimes the roots can appear on the outside, as if they were part of the trunk.

 Roots can take on many different forms: ramified in secondary roots, like hair, with a thick central root, and others.

The roots are an organ for absorption and are made up of various tissues: a protective **epidermis** on the outside, followed by the **cortex**, which serves as a food reserve, and inside, the germinating tissues of the **meristem** and the **conductive tissue** made up by the vascular bundles (phloem and xylem), which carry the crude sap.

THE PARTS OF A HIGHER PLANT

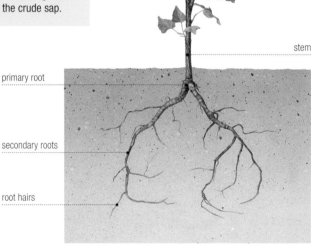

leaves

branch

stem

primary root

secondary roots

root hairs

FLOWERS

Flowers are the reproductive organs of higher plants. They can be unisexual, with either male or female elements, or hermaphroditic, with both types at the same time. The male reproductive structures are the **stamens**, made up of a filament that holds up a type of box known as an **anther**, which contains the **pollen**. The female reproductive structures are **pistils**, which consist of one or more ovaries that form an elongated chamber on a tube known as a **style**, which ends in a flattened structure, the **stigma**, which collects grains of pollen.

In this hibiscus, the stamens and anthers are clearly visible.

 Petals are leaves modified to serve a protective function and that generally are attractive. All of them together are known as the **corolla**.

 Sepals are leaves of scant beauty that are modified to serve a protective function. All of these leaves together are known as the **calyx**.

50

ANIMAL ORGANS

As with tissues, the organs of animals are more varied and complex than plant tissues. They can all be grouped into seven large categories: organs for digesting, breathing, blood circulation, excretion, control of body functions, reproduction, and sense perception.

The **digestive** organs introduce food into the animal's body (through the **mouth**), reduce it to a small size (using the mouth, crop, and **stomach**), and break it down into simpler substances to be absorbed as nutrients for the cells (using the **intestines** and related glands).

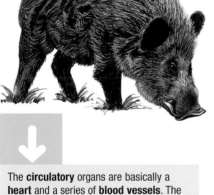

The sensory organs of an animal are the basis of the senses. In many animals, the sense of smell is essential in the search for food.

THE DIGESTIVE ORGANS OF A GRASSHOPPER

crop — gizzard — stomach — small intestine — esophagus — mouth — anus — rectum — salivary glands — gastric ceca — large intestine

The **respiratory** organs handle the exchange of gases between the interior of the animal and the exterior environment. Animals that live in the water breathe by means of **gills** or through their skin; those that breathe in the air use tubes known as **trachea** (in insects) or **lungs** (in vertebrates).

THE HEART OF A FROG

to the head — to the rest of the body — to the lungs and skin — left carotid artery — aortic arch — pulmocutaneous artery — right auricle — left auricle — ventricle — vena cava

There are various specialized organs for the movement of the body and controlling the organism's internal chemistry; these include the **muscles**, which are used for movement, and the **glands**, which control the internal conditions.

The **circulatory** organs are basically a **heart** and a series of **blood vessels**. The heart is a contractile organ that pumps the circulatory fluid (**blood** and **lymph**) through the network of vessels (**veins**, **arteries**, and **capillaries**) to bring it to the cells and supply them with oxygen and nutrients, and to remove the waste products.

A WOMAN'S REPRODUCTIVE ORGANS

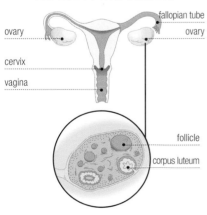

ovary — fallopian tube — ovary — cervix — vagina — follicle — corpus luteum

The reproductive organs have a double purpose: to produce male and female **sexual** cells (**gametes**) and then to unite them (through **fertilization**) to produce a **zygote** that will develop into a new individual. The male organs that produce gametes (**spermatozoids**) are the **testicles**; the female counterparts are the **ovaries**, which produce **ovules**. Some animals, such as mammals, reproduce through internal fertilization; for that purpose, the males are equipped with an organ (the **penis**) that introduces the spermatozoids into the female's body.

The **excretory** organs serve to eliminate harmful wastes from the body. In invertebrates, this is accomplished through tubes (**nephridia**), which connect the interior of the body with the exterior. In vertebrates, the organs used for this purpose are the **kidneys.**

The **sensory** organs are essential for an animal's life, since they keep it informed about the external and internal conditions so it can react in the most appropriate manner. An **internal receptor** controls the body temperature of mammals. External receptors include the **eyes** (for light stimuli), the **ears** (for sound stimuli), the **nose** (for chemical stimuli), and others.

THE PARTS OF A SQUID'S EYE

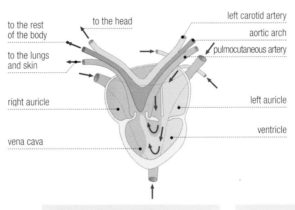

ciliar muscle — cornea — lens — pupil — iris — eyelid — cartilage — neurons — optic nerve — optical ganglia — retina

CELLULAR METABOLISM

Living beings are distinguished from lifeless, inert substances by the fact that they are capable of maintaining their own structural level; in other words, they maintain themselves throughout their life independently of external conditions. Metabolism consists of all the physical and chemical processes (exchanges of matter and energy) that a living being performs in order to maintain its structure and to reproduce.

ANABOLISM AND CATABOLISM

There are two basic reactions or processes in **metabolism**. First, there are those in which larger, more complex molecules are synthesized from simpler ones. This step, which is called **anabolism**, requires energy. On the other hand, cells also break down complex molecules in order to create smaller ones plus a certain amount of energy. The series of processes by which material is broken down to produce energy is known as **catabolism**.

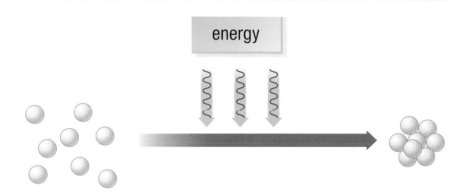

Anabolism: construction of organic material. For example, several glucose molecules make up a molecule of starch or cellulose.

Generally, the anabolic processing of a substance has a catabolic counterpart; however, these processes may proceed at different speeds, and they commonly take place in different parts of the cell.

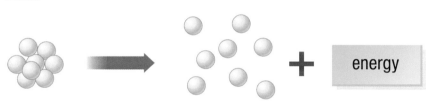

Catabolism: breakdown of organic material. For example, a molecule of starch is broken down into various glucose molecules.

METABOLIC PROCESSES

In order for one substance to be converted to another, it generally undergoes a series of transformations from the initial product to the final result. Each one of these transformations depends on a specific **enzyme**, in other words, on a catalyst that speeds up or slows down the reaction. The chain of reactions in sequence is known as the **metabolic process**. These processes exist in great numbers, and they take place inside living beings; furthermore, there commonly are relationships among them.

All living beings have molecules that help in the production of certain chemical reactions. These are **catalysts** or **enzymes**; without them, many reactions would be much slower, and the body would not function as efficiently.

BREAKDOWN OF ORGANIC MATERIAL THROUGH CELLULAR RESPIRATION

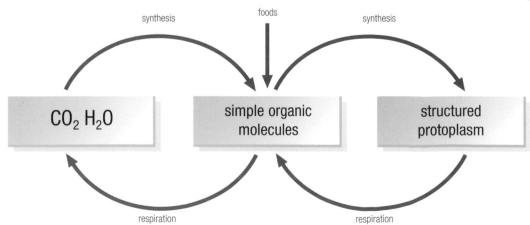

CHEMICAL BONDS AS A SOURCE OF ENERGY IN CELLS

There are many well-known energy sources, such as electrical energy, which makes it possible to light our houses, and fossil fuel energy (gasoline and other petroleum products), which makes it possible to run motors. As we have explained previously, the **human body** and the other **living beings** also require energy in order to function. In this case, the cells obtain energy by breaking certain **chemical bonds**. Animals and fungi extract the energy that is stored in the chemical bonds of the materials they take in as food. On the other hand, plants use the energy from solar light.

The energy for living beings is stored in a molecule known as **ATP** (adenosine triphosphate). When the cell needs energy, it breaks one of the phosphate bonds of this molecule and obtains energy (7.3 kcal/mol), producing a residue of one molecule of ADP (adenosine diphosphate) + P (phosphate).

To keep the supply of ATP from being used up, there is also a chemical reaction in which ATP is formed from ADP + P + energy. This reaction is carried out only in the mitochondria.

PHOTOSYNTHESIS

This is one of the most important metabolic processes of all the ones in existence, since it is the process by which **inorganic matter** is transformed into **organic matter**. It involves several steps; the first is one of the most important, since it captures the energy from solar light in order to manufacture **ATP**. In the second step, the energy of the ATP is used to construct sugars from the CO_2 in the air (inorganic matter). These sugars are organic material, and they are the principal parts of the body of photosynthetic organisms—that is, the **plants** and cyanophyceae.

Plants utilize the energy released from the sun's rays; animals, on the other hand, must metabolize it from foods (plants or other animals).

Photosynthesis is performed mainly in the leaves; but, in some plants, it also takes place in green stems. The process occurs inside the chloroplasts of the cells.

GROWTH AND DEVELOPMENT

From the time a living being is born until it reaches adulthood, it passes through a period of growth in which it undergoes a tremendous number of changes. In some cases, as with one-celled animals, this stage is very brief; but in others, such as trees and mammals, this growth period is comparatively long. The more highly evolved the plant or animal is, the longer its period of growth.

DEVELOPMENT

Development is the way in which a living creature attains maturity. There are two basic types of development: **direct** and **indirect**. In the first instance, the organism is born with an appearance and a body structure similar to an adult of the species; it increases in size and slightly modifies the function of certain organs. In the case of indirect development, the organism is born very different from the adult stage. To reach maturity, it must pass through a period of profound transformations known as **metamorphosis**.

Mammals develop directly, since, when the young are born, they have essentially the same shape as the adults; like this cow and the calf it is feeding, they are distinguished primarily by size.

THREE PHASES IN THE DEVELOPMENT OF AN INSECT

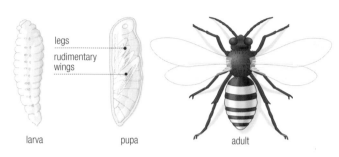

legs
rudimentary wings

larva   pupa   adult

Indirect development takes place with many insects and some amphibians, such as frogs.

In nature few animals are able to live fully their reproductive years and reach old age. This happens only in the case of humans, some primates, and large animals, such as elephants.

COMPARISON OF THE DEVELOPMENT OF EMBRYOS OF DIFFERENT ANIMALS

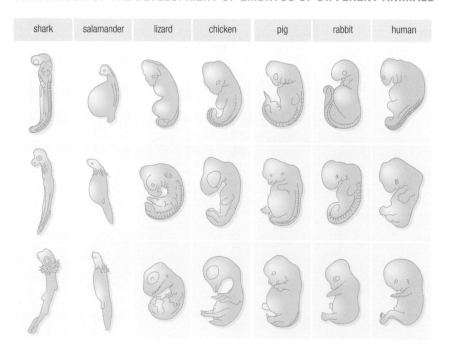

shark	salamander	lizard	chicken	pig	rabbit	human

FROM ZYGOTE TO ADULT

With multicelled animals that reproduce sexually, the starting point is the **zygote**, which results from the union of a **spermatozoid** and an **ovule**. From that moment on, the organism begins to develop, first by duplicating that initial cell and then dividing, in succession, the resulting cells until a multicelled mass of undefined shape is created. As growth continues, it resembles more and more the final adult form.

When an animal grows old, it becomes more vulnerable to attack by predators, to illness, and to harsh weather.

THE INFLUENCE OF GENES

All animals have a definite shape (all parrots have the beak in the same place and of comparable hardness and size); this shape remains the same among the individuals of the same **species**. Still, plants have an individual shape in every case, even though they may belong to the same species and have the same **genes**. This is because their growth depends heavily on **environmental** conditions—in other words, in a year of scant rain or low temperatures, plants may grow very little. However, in favorable times, they grow more and develop buds that remain dormant during unfavorable conditions.

In contrast to animal species, plant species grow and develop very differently.

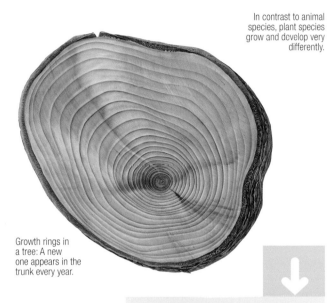

Growth rings in a tree: A new one appears in the trunk every year.

If the genes are altered or damaged, growth may be abnormal. In nature, there are instances of animals that grow with a short extremity, six toes, and so forth.

ILLNESSES DUE TO AN EXCESS OF A DEFICIENCY OF GROWTH HORMONE

Cause	Illness	Appearance
Deficiency during development	Dwarfism	Correct proportions, but small size
Excess during development	Gigantism	Correct proportions, but very large size
Excess after growth period	Acromegaly	The feet, hands, jaw, and other distal parts of the body experience excessive growth

Plants also have a growth hormone that is known as **auxin**.

THE GROWTH HORMONE

This is an indispensable molecule for the proper development of a human; it completes all the phases of development and produces healthy tissues and organs. A deficiency of growth hormone translates into incomplete growth, as with **dwarfism**. An excess, on the other hand, causes **gigantism**. Many other animals have corresponding growth hormones.

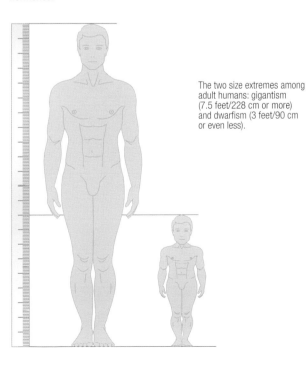

The two size extremes among adult humans: gigantism (7.5 feet/228 cm or more) and dwarfism (3 feet/90 cm or even less).

NUTRITION AND ENERGY

Living organisms take in material from the external environment and incorporate it into their bodies to carry out vital functions. The nutritional process involves several steps: capturing and ingesting of food, digestion, and excretion of unused or excess substances. In addition, nutrition provides organisms with the necessary material for producing the energy required for maintaining their structure.

INGESTION AND DIGESTION OF FOOD

In nature, there are many ways to get **food**, from the **roots** of plants to the mouth of animals, including taking in particles through the cellular membrane of one-celled animals. In any case, once the food has been ingested, it has to be digested; in other words, it has to be broken down into smaller particles that eventually become the pieces that the cells use to make their own molecules and structures.

When the particles to be ingested are quite large, endocytosis occurs, whereby the membrane surrounds the particle and invaginates it toward the inside, moving it toward the lysosomes where they are digested.

HOW A ONE-CELLED ORGANISM PHAGOCYTIZES FOOD

One-celled organisms have to take in food through their cellular membrane. This occurs in various ways, depending on the type of food.

capture

pseudopod

alimentary vacuole

LIVING ORGANISMS CLASSIFIED BY FOOD SOURCE

How living organisms get their food.

• **Autotrophic** organisms take nourishment from inorganic matter. **Photoautotrophs** use the energy from the sun's light to carry out photosynthesis and change inorganic substances to organic ones; **chemoautotrophic** organisms use the energy of chemical bonds in certain inorganic compounds.

• **Heterotrophic** organisms take their nourishment from organic material. Heterotrophs can get their food in several different ways.

  • When the basis of the diet is inert organic remains (dead bodies), the organism is said to be a **saprophyte**.

  • When the organisms feed on living beings that they capture, they are said to be **biophagous**.

  • Organisms that feed on the body fluids of a living creature and cause it harm; they are termed **parasites**.

  • Finally, there are creatures that take advantage of other living organisms but in a mutually beneficial relationship; these are **symbiotic** organisms.

1. photoautotroph (tree)
2. chemoautotroph (bacteria)
3. saprophytic heterotroph (fungi)
4. biophagous heterotroph (dog)
5. parasitic heterotroph (tick)
6. symbiotic heterotroph (lichen)

FOODS

Foods are the substances that the body utilizes to carry out its vital functions. There are three main types that must be consumed in adequate proportions. First, there are the **carbohydrates** (sugars and starches), which are found mainly in plants (vegetables, grains, and fruits). Second, there are **fats**, which are found in foods of animal origin (milk, meat, and fish). Third, there are **proteins**, which are present mainly in foods of animal origin (eggs, milk, meat, and fish); however, some dried fruits and vegetables are also rich in protein.

Vitamins are essential to life. Since our body cannot manufacture them, we can get them only through the foods that we ingest.

An assortment of foods consumed by humans; they contain carbohydrates, fats, and proteins.

RESERVE SUBSTANCES IN ANIMALS

In animals, the main energy **reserve tissue** is fat. When the body needs energy, it triggers a series of chemical reactions that oxidize fat and convert it to **energy**. In one gram of fat, there are 9.4 kcal. Energy can also be obtained by breaking down glucosides and proteins but that produces only 4.1 kcal per gram.

Many animals spend a major part of the spring and summer gorging themselves to accumulate a thick layer of body fat so they can survive a tough winter when it is difficult to find food.

In order to be ready for winter, during which it hibernates in its den, the bear needs to build up fat reserves.

GAS EXCHANGE

All living organisms exchange gases with their environment through a process known as **respiration**. The ultimate goal of respiration is to introduce molecules that aid in breaking down organic matter and thus release the energy from the molecular bonds that are broken. There are two main types of respiration: aerobic, in which the organisms consume oxygen, and anaerobic, which takes place in the absence of oxygen.

AEROBIC RESPIRATION

In this process, the organism takes in the **oxygen** it needs to live and gives off the **carbon dioxide** produced in the cells and which is toxic in excess quantities. Higher land animals breathe using **lungs**, and insects use **trachea**; most aquatic organisms obtain oxygen from the water, generally through **branchiae** or **gills**. Some animals that live in very moist areas (such as worms, amphibians, and eels) may breathe through their skin, and one-celled organisms breathe through their cell membrane.

Plants breathe, too, taking in O_2 and releasing CO_2. However, during the day or in the presence of light, they also carry out photosynthesis, taking in CO_2 and releasing O_2.

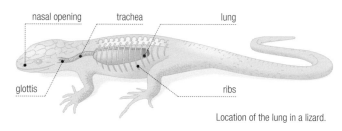
nasal opening   trachea   lung

glottis   ribs

Location of the lung in a lizard.

In all animals, the oxygen enters the circulatory system where the hemoglobin carries it to the cells.

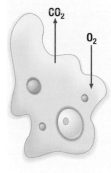
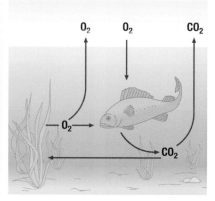
Two different ways of breathing. Left, an amoeba; right, a fish.

ANAEROBIC RESPIRATION

A few organisms live in environments where there is no oxygen; they are mainly **bacteria** that get their energy through a process in which oxygen plays no part.

Some of these organisms are strict **anaerobes**, which means that they can live only in places where there is no oxygen, since this element is toxic to them. However, there are other organisms (some bacteria and fungi) that are **optional anaerobes**; this means that they can live in environments that have oxygen or in **anoxic** ones (which have no oxygen).

At the bottom of lakes and swamps where the water coming in is very polluted, the dissolved oxygen is consumed very quickly until there is no more.

When the water of the ocean, a river, or a lake is polluted, fish cannot breathe or take in nutrients, so they die.

FERMENTATION

This is, in fact, the same process as anaerobic respiration, but, in general, fermentation occurs when the organic matter is not entirely broken down, but rather the result is various residual organic compounds (generally acids). This process, which is carried out by microorganisms, is very important to people because it makes possible the transformation of many organic materials into other very appealing ones, such as yogurt, cheese, beer, and wine.

At some time, we all have noticed the foul odor of decomposing meat or a rotten egg. This smell is due to the compounds that result from the fermentation of the proteins carried out by certain bacteria.

There are several types of fermentation, and each one yields different products: lactic (cheese and yogurt), alcohol (wine and beer), and acetic (vinegar).

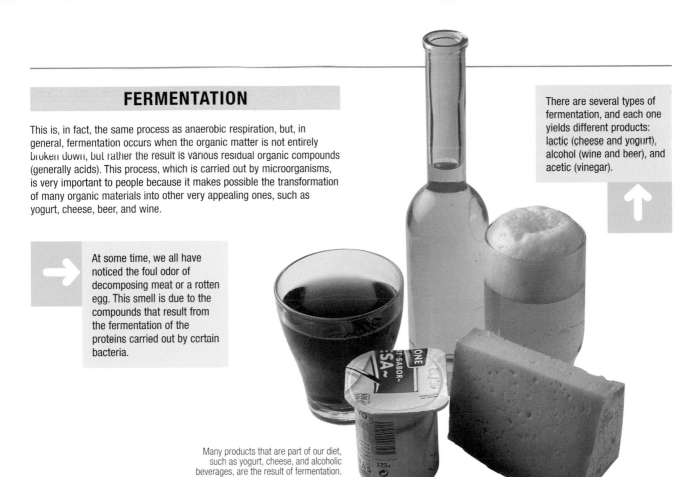

Many products that are part of our diet, such as yogurt, cheese, and alcoholic beverages, are the result of fermentation.

THE AIR WE BREATHE

The Earth's atmosphere contains about 21% oxygen at sea level. As altitude increases, the surrounding pressure decreases, and there is an attendant decrease in the oxygen concentration. As a result, the greater the altitude, the harder it is to breathe. The lack of oxygen becomes noticeable starting around 10,000 feet (3,000 meters), and every movement requires greater exertion.

The combustion of any substance uses up oxygen from the air. In the case of gasoline or diesel combustion, carbon monoxide (CO) is also produced. This gas is very toxic, since the hemoglobin in the blood absorbs it more easily than oxygen and that can lead to death.

One of the problems encountered by mountain climbers on high peaks is the thinness of the air; the oxygen content decreases with altitude, so starting at around 16,500 feet (5,000 meters), they have to use oxygen masks.

TRANSPORT OF MATTER IN LIVING ORGANISMS

There is a great deal of bustle inside living beings. The nutritive elements, the cells, hormones, gases, and other substances are in continual movement from one place to another so they can be used in the places where they are most needed, and they always move through the channels that are made for the purpose. In addition, they rarely travel solo, but rather they piggyback onto transporting molecules that bring them to their final destination.

SAP

Plants have a sort of circulatory system that is made up of a series of tubes that run inside the stalks and the leaves; this is where the liquid known as **sap** flows. The water and the mineral salts drawn in by the roots travel toward the leaves through special tubes (**xylem**) in the form of a liquid that is known as **crude sap**. In the leaves, **photosynthesis** produces glucose and other molecules necessary for the formation of tissues. These molecules travel through other ducts (**phloem**) to the rest of the plant, and the liquid is known as **processed sap**.

Plants absorb energy from the sun and transform the crude sap (a mixture of water, glucose, and other organic substances).

THE CIRCULATORY SYSTEM

Animals transport nutrients, gases, and wastes for excretion in their **blood**. From the intestine, the digested nutrients, converted into small molecules, pass into the bloodstream, which carries them to the tissues where they are needed. Gases from respiration (**oxygen** and **carbon dioxide**) combine with the **hemoglobin**, which transports the oxygen from the lungs to the tissues and the carbon dioxide from the tissues back to the lungs. The waste products are carried from the tissues, where they are produced, to the kidneys, where they are eliminated.

HOW A PLANT GETS ITS FOOD

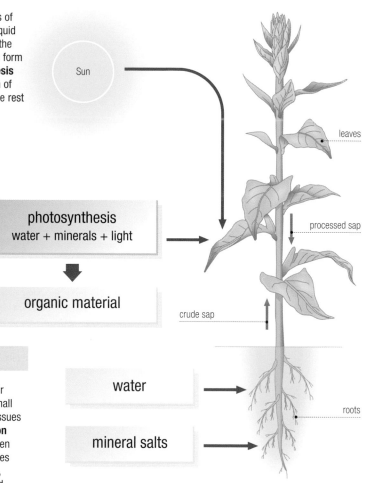

Sun

leaves

photosynthesis
water + minerals + light

organic material

processed sap

crude sap

water

mineral salts

roots

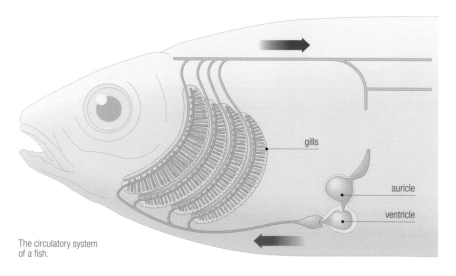

gills

auricle

ventricle

The circulatory system of a fish.

The motive force for the blood comes from the heart, which constantly maintains circulation. Invertebrates may have several hearts, but vertebrates have just one; however, it is much more highly refined.

HOW LIVING ORGANISMS FUNCTION

60

THE PASSAGE OF SUBSTANCES THROUGH CELL MEMBRANES

The membrane that encases cells is **semipermeable**—in other words, it lets some substances pass through but not others. Small molecules, such as oxygen, carbon dioxide, salts, and others, pass from one side to the other through some channels made up of proteins known as **permease**, which enables them to pass through. This movement is **passive**—that is, it requires no energy consumption and is carried out by **diffusion** because of the difference in concentrations. When a cell wants to expel something toward the exterior and the concentration in the surrounding environment is greater than inside the cell, it resorts to using another type of protein to pump the element out by using energy; this is **active transport**.

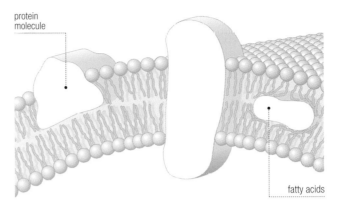

protein molecule

fatty acids

Fats pass through the membrane because the membrane contains fatty acids.

BLOOD TRANSPORT

(in ml/min) in different organs of the human body at rest and during vigorous exercise

Organ	Rest	Exercise
heart	250	750
muscles	1,000	12,500
brain	750	750
kidneys	1.200	600
skin	400	1,900
intestines	1,400	600

When a cell has to take in large particles, it resorts to **pinocytosis** or **phagocytosis**, which involve invaginating the membrane and surrounding the particle or liquid.

EXCRETION

All chemical reactions that take place inside living beings produce waste substances. Many of them are toxic, and some simply do not fit inside the cells and must be excreted to the outside. **One-celled** organisms eliminate wastes by means of **vacuoles**. **Multicelled** organisms accomplish this through their **circulatory** and **excretory** systems. Plants accumulate these substances in their tissues, but animals expel them to the outside in the process of **excretion**. Harmful gases, such as carbon dioxide, are eliminated through respiration, and harmful substances, by means of the circulatory system and special organs such as the **kidneys**.

Ingested foods are processed and absorbed in the intestine. The blood carries all these substances through the body to the organs that need them.

Solid wastes produced by digestion are excreted in the form of feces. The toxic substances that circulate in the blood pass through the kidneys where they are converted to urine, which is excreted.

GENERAL DIAGRAM OF EXCRETION IN LIVING BEINGS

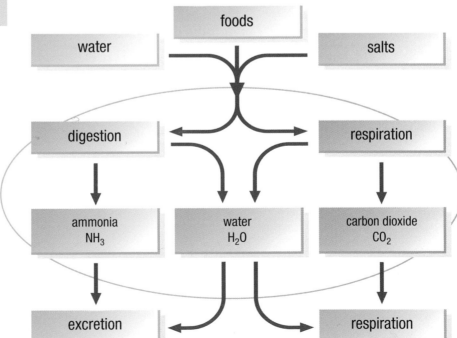

foods

water

salts

digestion

respiration

ammonia
NH_3

water
H_2O

carbon dioxide
CO_2

excretion

respiration

HOMEOSTASIS, THE EQUILIBRIUM OF LIFE

Environmental conditions can limit the life of living beings. But, inside the body of every living creature there is a process of physical and chemical processes that make it possible to maintain a more or less constant internal environment independently of the exterior. This series of reactions that make it possible to achieve internal equilibrium is known as homeostasis; it is essential to maintaining life.

THE INTERNAL ENVIRONMENT

Living creatures need very specific conditions so that their organs and cells can function properly. This set of conditions is what is known as the **internal environment**, and it usually involves a liquid containing **mineral salts** and **proteins** in the proper concentrations. One example that helps in understanding the importance of **homeostasis** is the sweating that occurs in response to elevated temperature inside the body. When the temperature norms are exceeded, through exercise or environmental heat inside the body, it reacts by sweating. This serves to remove excess heat to the outside and keep the internal temperature within proper limits so the body can function properly.

Tissues need a consistent degree of **humidity**, **pH**, **temperature**, and so forth—in other words, an environment that provides the proper conditions for their functioning.

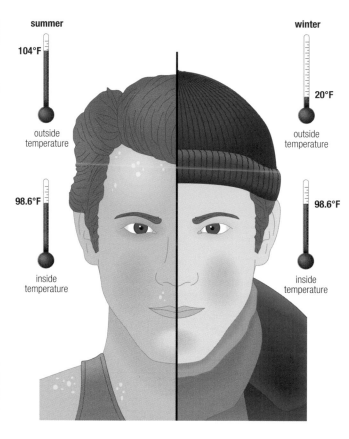

summer

104°F

outside temperature

98.6°F

inside temperature

winter

20°F

outside temperature

98.6°F

inside temperature

Whether it is hot or cold outside, the body temperature is kept constant by means of homeostasis.

ALL LIVING CREATURES HAVE AN INTERNAL ENVIRONMENT

In effect, all creatures require an internal environment that provides the necessary balance for carrying out all the required life functions: breathing, taking in food, growing, reproducing, and so forth. However, not all living beings have the same degree of complexity, so they need a fairly precise equilibrium.

Plants are living beings that generally can withstand great temperature variations; their internal environment does not change, and they merely speed up or slow down their activity based on the temperature. The same thing happens with **cold-blooded** animals—in other words, with invertebrates, fish, amphibians, and reptiles. Birds and mammals, which are **warm-blooded**, have a very complex internal environment, which they maintain between very narrow limits—in other words, the process of homeostasis is very precise.

An invertebrate's metabolism has more regulatory processes than that of a plant but less than that of a vertebrate; as a result, its homeostasis falls between the two extremes.

active in summer, inactive in winter

lives in summer, dies in winter

active all year

How organisms live in the different seasons.

SALINITY

Aquatic organisms generally are highly dependent on the **salinity** of the water in which they live. Most freshwater fishes die when they are placed into salt water. This happens because their body is not equipped with the right mechanisms for expelling the excess salt that gets in through the mouth and the gills, elevating the salt level in the internal environment and upsetting the equilibrium. The same thing happens in reverse, when saltwater fish are put into freshwater, they die; the cells in their tissues burst because of the excess intake of water in reaction to the excess of salt compared to the environment.

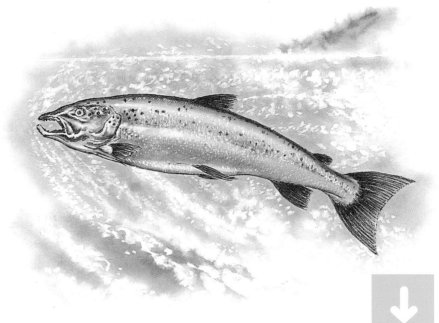

Some fish can live in either salt water or freshwater. They are said to be euryhaline—*euri* means *great*, and *haline*, *salinity*.

Only a few fish, such as the salmon and the eel, are adapted to both freshwater and salt water.

Eels and salmon are euryhaline fish that spend part of their life in freshwater and part in salt water.

THE INTERNAL EQUILIBRIUM CAN INFLUENCE THE EXTERNAL SHAPE

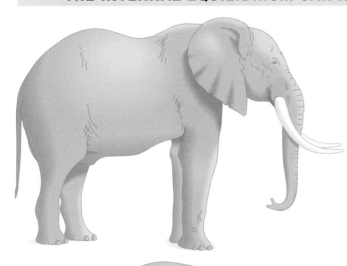

African elephants and **Asian elephants** are very similar animals in almost every regard, but their most distinguishing trait is the size of their ears. This size difference is not mere happenstance, but rather an adaptation to the environment in which they live. In Africa, they live on the savannas, where the climate is extremely hot; in Asia, the elephants live in areas that have abundant forests. As a result, the body of African elephants tends to build up more heat. To shed this excess heat, they use their ears, which contain a great number of **blood vessels** separated from the outside by a thin layer of skin; the more blood vessels there are, the more efficient the cooling system. An exception in Africa is the **woods elephant**, which also has smaller ears than the elephants that live on the plains.

Many other desert animals have abnormally large ears to get rid of body heat; one example is the desert fox.

A desert fox, with its large ears, inside its den.

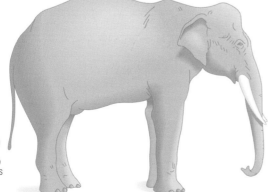

The African elephant (above) is a little larger than the Asian elephant (below); it also has larger ears and tusks.

PLANT IRRITABILITY

The interaction of living beings with their environment and the other organisms that live around them is one of the most important features of their lives. Irritability is the term that is used to describe the response of an organism to a change in its environment that may affect it. When the change causes the organism to reduce its activity, it is said to cause **inhibition**; on the other hand, if it increases activity and a heightened use of energy, it is known as **excitation**.

STIMULUS PERCEPTION IN PLANTS

One of the main differences between plants and animals is that the former have no **nervous system**. Just the same, that does not mean that they cannot receive stimuli from the outside and react to them. Plants have **receptors** located inside them (sometimes in the cells), which are capable of receiving physical stimuli (such as light, gravity, temperature, and others), as well as chemical stimuli (including the presence of water and nutrients). Once they pick up the stimulus, they generate **hormones** that travel to the part of the plant where the response is needed—in other words, they act as messengers.

PLANT STIMULI

spring (the day grows longer)

summer

autumn (the day grows shorter)

winter

Plants that grow in areas that experience climatic variations throughout the year are very seasonal. When the day starts to lengthen, growth and flowering are induced. When the hours of light grow fewer, they lose their leaves and reduce their activity.

Plant hormones are known as **phytohormones**; they correspond to the hormones that are present in animals.

MOVEMENT IN PLANTS

Even though plants cannot move of their own will the way animals do, they are capable of certain active movements in response to external stimuli. There are several types of movement, including **taxis**, **tropisms**, and **nastic movements**. In addition, certain parts of plants (such as seeds, pollen, and others) are capable of traveling to distant places through passive movement by such means as wind or water currents and sticking to animals.

There are some plants that move on their own without any external stimulus. This type of movement is known as **nutation**. An example is climbing vines that grow and curve in some direction in search of a support point.

Taxis	Property of lower plants and reproductive elements (gametes). This involves a cell's change of direction in response to external stimuli. This response can be positive, when it is directed toward the stimulus, or negative, when it goes in the opposite direction.
Tropisms	Growth movement in a certain direction in response to a stimulus. The tropism is positive when the plant moves toward the stimulus (such as a stem and leaves that grow toward light) and negative when the plant moves in the opposite direction (as with roots that grow away from light). There are several types: • Phototropism: when the stimulus is light. • Geotropism: when the stimulus is the force of gravity. • Chemotropism: when the stimulus is chemical (such as the presence of moisture). • Haptotropism: when the stimulus is a solid body (as in the case of climbing vines that grow onto a substrate and cover it).
Nastic movements	Movements of a plant in response to an external stimulus. There are several types: • Haptonasty: when the stimulus is contact (as with carnivorous plants that close their jaws when an insect lands inside them). • Photonasty: when the stimulus is light (for example, many flowers open in the presence of light and close at night, or vice versa). • Thermonasty: when the stimulus is the temperature. • Chemonasty: when the stimulus is chemical.

EXPERIMENTS

1. In Search of Light

We simultaneously plant one seed in a normal flowerpot (A) and another seed in a pot placed inside a shoe box (B) with a small opening through which light can come in where we want it. After several days (provided that we water and care for the plants), we see that the first plant grows straight up—normally—and the other one grows toward the hole (or more precisely, toward the light source).

A

B

Light is one of the main stimuli for plants, since it provides the energy they need to carry out photosynthesis.

2. The Law of Gravity

Place a seed onto a damp cloth (C) and allow it to germinate. After that happens, plant it with the root facing upward and the stem pointed down. After a few days we will see that the growth direction of these parts has changed (D). The root will bend downward (following the force of gravity) and the stem upward (opposite the force of gravity and toward the light).

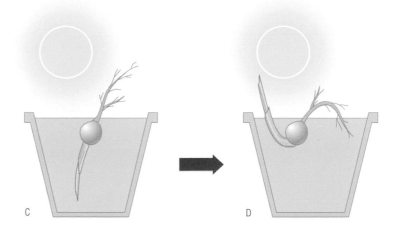

C

D

A GREAT INVENTION: THE NERVOUS SYSTEM

The nervous system, which is present only in multi-celled animals, is a very sophisticated system that allows the organism to be aware of environmental changes so that it can react to them appropriately in every instance. The nervous system also processes internal information about the organism's body—temperature, for example.

WHAT IS THE NERVOUS SYSTEM LIKE?

The nervous system is made up of a series of interconnected cells. They are linked in a chain sequence, and these chains are linked together to form a complicated network. **Sensory cells** are located at one end of these chains; these are the cells that pick up the **stimulus**, such as the ones in the eyes that detect light differences, the ones in the skin that perceive temperature changes, and so forth. At the other extremity is the **brain** or the **central ganglion**, which is where the information is processed. When the brain decides the response it intends to make to the stimulus, it sends information to the body's muscles or organs (to the motor cells) so that they can take appropriate action.

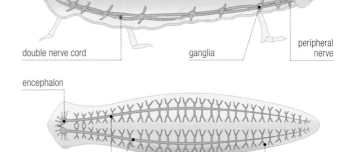

encephalon

double nerve cord   ganglia   peripheral nerve

encephalon

double nerve cord   peripheral nerve

The nervous systems of a grasshopper (upper) and a planarian (lower).

THE STRUCTURE OF A MOTOR NEURON

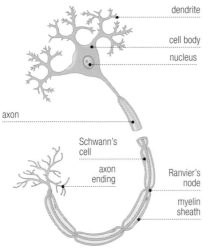

dendrite
cell body
nucleus
axon
Schwann's cell
axon ending
Ranvier's node
myelin sheath

Information travels through the nervous system in the form of electrical impulses.

Dendrites are the nerve fibers through which nerve impulses enter the neuron. The **axons** are the nerve fibers through which the impulse is sent out.

Nerve cells are known as **neurons**. They consist of a cell body, the **soma**, and extensions that are the **nerve fibers**.

STIMULI

These are the environmental conditions that can be perceived through the sensory organs. For example, when a dog sees a cat in the distance, it receives a visual stimulus; when its owner touches it, a tactile stimulus; when it hears a sound, an auditory stimulus; when it encounters the markings of another dog, an olfactory stimulus; and when it eats, a taste stimulus. On the other hand, there may be a variation in the pH of its drinking water that it cannot perceive, since it has no sensory organs that can detect pH; however, the microorganisms that live in that water may be capable of detecting the change.

HOW STIMULI WORK IN ANIMALS

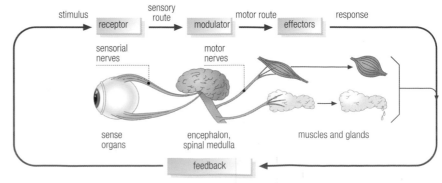

stimulus → receptor → sensory route → modulator → motor route → effectors → response

sensorial nerves   motor nerves

sense organs   encephalon, spinal medulla   muscles and glands

feedback

Dogs and many other mammals have a highly developed sense of smell. That is due, in part, to their long nasal passages, which contain a greater number of cells that are more sensitive to smell than we humans have.

SPECIAL SENSES

Some animals have developed a capacity to pick up stimuli that others cannot. This is one means of adapting to the environment and to a way of living. Thus, **bats**, for example, fly in the dark, constantly sending out sonar signals that rebound off objects. When this sound strikes a branch, for example, it produces an echo that the bat can detect and thus avoid the branch. The sounds that bats emit are **ultrasound**, which is undetectable to humans and many other animals.

Whales also can communicate with one another over long distances by means of **ultrasound** that they send through the water. Their nervous system has developed a very acute sense of hearing that enables them to survive in an environment as vast as the ocean.

Bats can fly around obstacles thanks to their sophisticated "radar" system.

BEHAVIOR

The way an animal behaves is the product of its nervous system. In principle, lower animals exhibit a limited range of behavior, since they act **instinctively**. As the degree of complexity increases, the behavior of the organisms likewise becomes more highly evolved. At the top of this scale are humans, whose behavior normally does not depend on instincts; rather, they usually react to environmental stimuli influenced by behavior. For example, they drink liquids at a meeting or a party even though they are not thirsty.

Etiology (from the Greek *ethos*, meaning *custom* and *logos*, treatise) is the science that studies the behavior of animal species; its goal is to explain how all living creatures function in relationship with their environment.

The more highly evolved an animal is, the more complex its nervous system and the larger its brain. The larger the brain, the more complex the animal's conduct, since it can store memories and experiences that will provide future answers to the same stimuli.

The courtship of the three-spined stickleback is ritualized; it is an instinct triggered by the hormone system.

PASSIVE AND ACTIVE MOVEMENT

As we have seen in previous chapters, animals have a great capacity for moving about. However, there are many other types of movement that may not be as easy to detect, such as the heartbeat, the movement of the intestines to move the food along, and others. Active movement is done voluntarily, and passive movement is done by the body to perform its vital functions and is not subject to control.

THE AUTONOMOUS NERVOUS SYSTEM CONTROLS ACTIVE MOVEMENT

When an animal moves voluntarily, it uses a part of its **nervous system** that is referred to as **autonomous**. This contains two parts that can be distinguished: the **central nervous system**, made up of the brain and the spinal medulla, and the **peripheral nervous system**, which consists of the distal or sensorial nerves, which pick up the stimuli, and the motor nerves, where the muscles are activated.

One example of active movement involves capturing food. A chameleon, for example, sees an insect flying nearby and follows it with its eyes (it perceives the visual stimuli with the sensitive peripheral nerves of its **eyes** and transmits this information to the **brain**). When the insect is close enough, the brain sends an order to the muscles in the tongue to unroll and shoot out at the insect, capture it, and bring it into the mouth.

The chameleon suddenly shoots its tongue out at the insect, which sticks to it.

 The autonomous nervous system (ANS) is responsible for the ability to **react** to a stimulus.

THE VEGETATIVE NERVOUS SYSTEM CONTROLS PASSIVE MOVEMENT

There are many movements that animals carry out unconsciously, such as **respiration**. In fact, we do not continually have to think about breathing; we do it without realizing it. The part of the nervous system that is responsible for coordinating this type of movement is known as the **vegetative** nervous system. The brain still gives the orders and transmits them through the **nerves** until they reach the muscles. When the result is a stimulation of a process (such as increasing the heartbeat or concentrating ions in the blood), we say that the **sympathetic nervous system** has been activated; on the other hand, when the result is the opposite (because the process is inhibited), the **parasympathetic nervous system** comes into play.

 The vegetative nervous system (VNS) is responsible for the ability to **act** on organic processes and the condition of the animal.

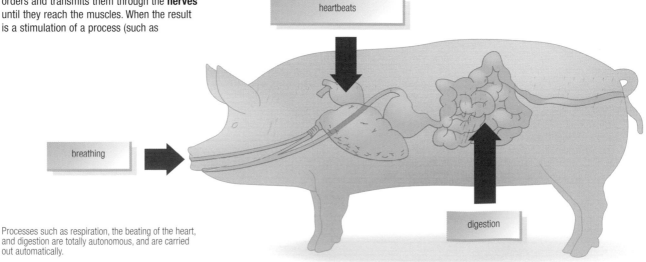

heartbeats

breathing

digestion

Processes such as respiration, the beating of the heart, and digestion are totally autonomous, and are carried out automatically.

MUSCLE TISSUE IS RESPONSIBLE FOR MOVEMENT

Except for one-celled organisms, sponges, and cnidarians, all animals have a well-developed **muscular system**. The muscles respond to orders delivered to them by the nerves, and they contract or lengthen to produce movement in a part of the body. **Smooth muscles** are used in passive movement, and **striated muscles** come into play in active movement.

 Striated muscles are given that name because they are made up of a tissue that looks striped because light and dark bands of muscle alternate; smooth muscles do not have these striations.

 Plants also move, although generally so slowly that the movement is barely perceptible.

MUSCLE TISSUE

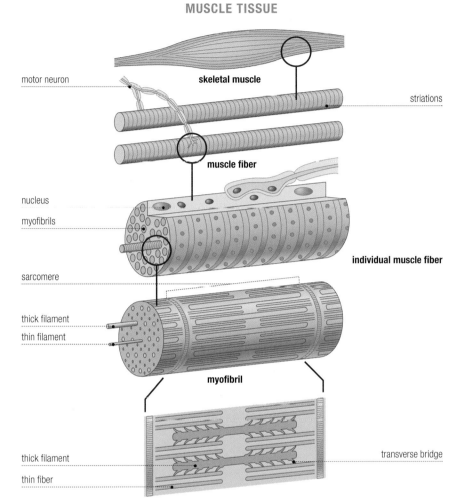

motor neuron

skeletal muscle

striations

muscle fiber

nucleus

myofibrils

individual muscle fiber

sarcomere

thick filament

thin filament

myofibril

thick filament

thin fiber

transverse bridge

CELLULAR MOVEMENT

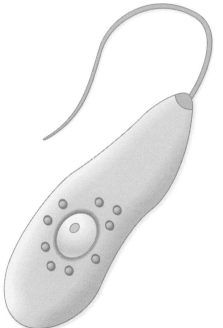

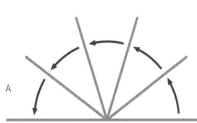

A

B

Most of the cells that make up a tissue are immobile, but there is a lot of movement inside them, including the passage of substances into and out of the cell through the membrane, which has specific channels for each substance.

On the other hand, some special cells of multicelled animals, such as **spermatozoids** used in reproduction, as well as many one-celled organisms, have the ability to move on their own. In order to move, they use either **cilia** or **flagella**, or else they creep along the substrate by means of amoeboid movements, in other words, by putting forth **pseudopods** the way amoebas do.

The movement of the flagellum of a protozoan (left) has two stages: (A) the advancement stage and (B) the recovery stage.

ANIMAL BEHAVIOR

Behavior is the set of an organism's reactions to the world that surrounds it. In response to a stimulus such as the sunrise, bats avoid it by hiding in their caves, but reptiles come out to warm up in the sun's rays. Every animal species has its own way of behaving, and, in most cases, this behavior is encoded in the genes and is one of the main tools for guaranteeing survival.

BEHAVIOR AND THE NERVOUS SYSTEM

Protozoa, which are one-celled creatures, respond to stimuli from the environment according to **chemical** or **physical laws**; for example, when they encounter an excessively salty area, they turn around and retreat in search of a more appropriate area. As the degree of complexity in the **nervous system** increases, behavior becomes more and more elaborate. Worms, for example, can hide in the ground or come out of it in response to the amount of environmental moisture or heat; spiders spin webs to capture their victims; and birds can communicate with one another with their songs. Humans are located at the top of the range; their behavior is very complex, since it involves the process known as **thought**.

The more complex the nervous system, the more abilities the creature can develop.

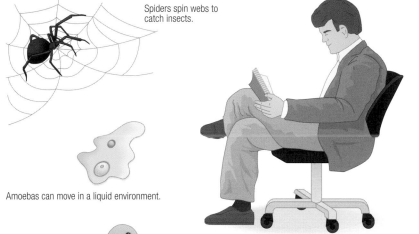

Spiders spin webs to catch insects.

Amoebas can move in a liquid environment.

Birds use their songs to communicate with one another.

Humans can think.

The science that focuses on animal behavior is known as **etiology**; the one that is devoted to human behavior is **psychology**.

LEVELS OF BEHAVIOR

Animals exhibit two basic types of conduct. In some cases, they perform actions automatically, always with the same response to the same stimulus; this is **innate behavior** (taxis, reflex actions, and instinct). Another level of behavior is the one referred to as **acquired**, which is built up on the basis of life experiences (learning and reasoning). Acquired behavior is the province of higher animals, especially vertebrates; however, there are some invertebrates, such as octopuses, that have a high level of intelligence.

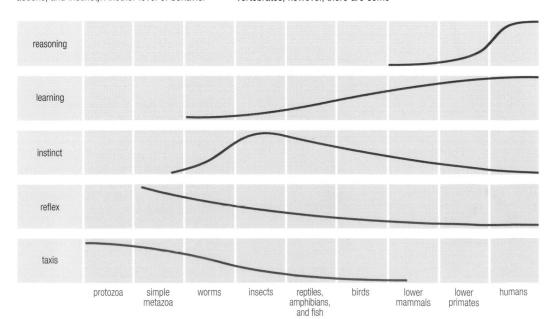

Relative importance of the different types of behavior.

reasoning									
learning									
instinct									
reflex									
taxis									
protozoa	simple metazoa	worms	insects	reptiles, amphibians, and fish	birds	lower mammals	lower primates	humans	

Another level of behavior is social—in other words, the reactions that take place in the presence of individuals of the same species. This is especially important in some animals, such as the primates.

INSTINCT

Instinct is the behavior used by animals in reaction to certain stimuli and which is very helpful to them in survival. This involves more complex patterns of **behavior** than mere **reflexes**, which function in the same way, always producing the same response; however, in higher animals, these actions can be accompanied and modified by other behaviors. One of the basic instincts of most animals is self-preservation. The response can vary greatly, so in response to an attack by a **predator**, in some cases, the response is **flight**; in others, the animal remains completely immobile so the predator cannot see it; and in others cases, the animal faces the predator to defend itself. Other instinctive behaviors target **reproduction**; **caring for the offspring**, in the case of birds and mammals; and **migration**, in both vertebrates and invertebrates.

Small eels are born in the Sargasso Sea. Not long thereafter they undertake a long journey to the rivers of Europe and North America. There they grow and develop, and when it is time to reproduce, their instinctive mechanism kicks in once again, and they start the voyage back to the place where they were born to lay their eggs.

Other well-known migratory species are the monarch butterfly, salmon, reindeer, gnu, storks, and swallows.

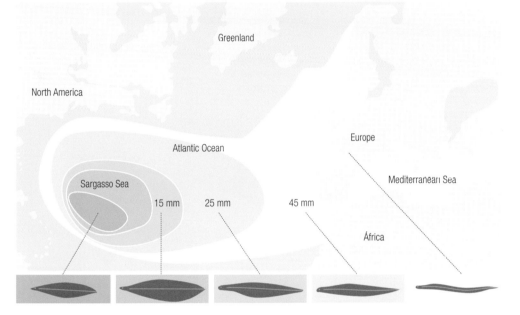

Greenland

North America

Atlantic Ocean

Europe

Mediterranean Sea

Sargasso Sea

15 mm   25 mm   45 mm

África

LEARNING

Animals that have a complex **nervous system** are capable of learning from life experiences. For example, seagulls have observed that a great number of fish with no economic value are thrown into the sea from fishing vessels, and, over time, they have learned to follow these ships to get a lot of food with little effort. Following ships is not an innate behavior with these creatures but rather an acquired one based on experience.

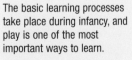

The basic learning processes take place during infancy, and play is one of the most important ways to learn.

Learning looks like a game for lion cubs, but it is really a continual development of skills that will enable them to defend themselves and get food on their own.

The more highly evolved the psyche of an animal, the longer its learning period. So infancy lasts for several years in the primates, including humans.

ASEXUAL REPRODUCTION

Since all living creatures have a limited life span, it is essential that they multiply in order to perpetuate the species. Reproduction is the phenomenon by which existing organisms create new individuals from themselves. One way of reproducing is asexual, which involves no exchange of genetic material. There is no exchange of genes, so the descendants are always identical to the progenitor, since their genetic material is always the same.

TYPES OF ASEXUAL REPRODUCTION IN ANIMALS

In the animal world, there are four basic types of asexual reproduction. One-celled organisms generally reproduce by means of **bipartition**: the mother cell divides into two equal parts to form two identical individuals. In other cases, several new individuals can spring from the mother cell; this is called **multipartition** or **multiple excision**. **Gemmation** is a process involving the formation of a series of body extensions or offshoots, which develop into new individuals. Finally, there is **fragmentation**, which occurs in multicelled organisms; it involves dividing an individual into two or more fragments that develop into new creatures.

ASEXUAL REPRODUCTION IN LOWER ORGANISMS

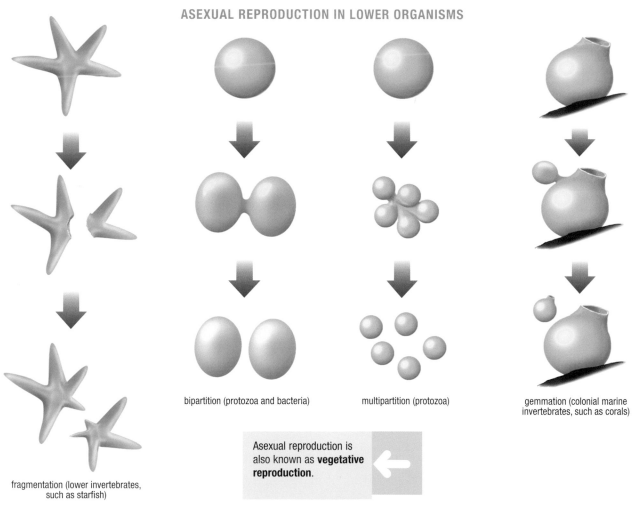

bipartition (protozoa and bacteria)

multipartition (protozoa)

gemmation (colonial marine invertebrates, such as corals)

fragmentation (lower invertebrates, such as starfish)

Asexual reproduction is also known as **vegetative reproduction**.

ASEXUAL REPRODUCTION IN PLANTS

Plants have a life cycle in which asexual and sexual phases alternate. In the asexual phase, a mother cell divides to form a series of cells on the basis of which new plants develop. These cells are called **spores**; they have a protective coating to keep them from drying before they find appropriate conditions for germination.

Plants also reproduce by gemmation or fragmentation (by forming propagules, tubers, rhizomes, stolons, bulbs, and so forth).

Bacteria, fungi, and algae also produce spores for asexual reproduction.

GEMMATION

This is a process of asexual reproduction that can characterize both one-celled and multicelled organisms. In the former case, a projection is formed from a part of the protoplasm, and this occurs generally in yeasts. In the case of multicelled organisms, such as sponges and coelenterates, the projections are made up of a series of cells. The projection can either detach itself from the progenitor's body or remain united with it to form a colony.

Asexual reproduction does not allow for evolution, since there are no differences between progenitors and descendants that could improve environmental adaptation.

Coral reefs are an example of gemmation, since the new individuals remain attached to the progenitors and make up a very large mass.

THE ALTERNATION OF GENERATIONS

In lower organisms, the only way to reproduce is asexually; however, as the plants and animals become more complex, they begin to reproduce sexually. In such cases, sometimes they reproduce by one means and, at other times, by the other, according to a cycle known as **alternation of generations**. This alternating is very advantageous, for, on the one hand, it allows for quick and simple reproduction, and, on the other, there are phases in which genes can be exchanged, thereby making **evolution** possible.

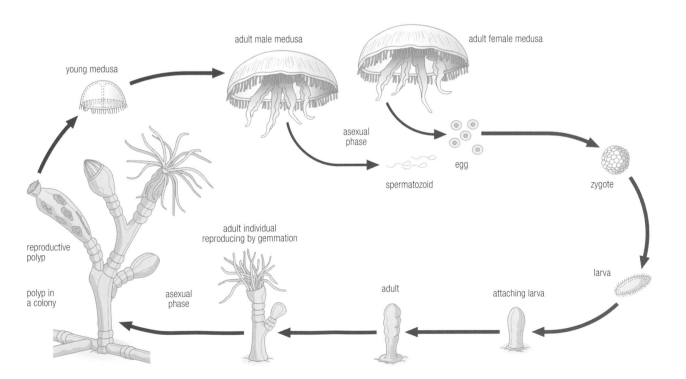

73

SEXUAL REPRODUCTION IN PLANTS

Sexual reproduction involves an exchange of genetic material between two individuals. Each one of them produces specialized cells called gametes; when they unite, they form a zygote, which will develop into a new organism. Plants are typical organisms with a life cycle in which asexual and sexual reproduction alternate.

GAMETES

Gametes are cells that contain only half the genetic information and are known as **haploid** cells. They are the result of a division process called **meiosis**, in which four haploid cells, each with one set of chromosomes, are formed from one cell with two sets of chromosomes (known as a **diploid** cell). Each of the cells produced has a different name, and its sex is also a factor. Thus, male gametes are **pollen grains**, and female ones are **ovules.**

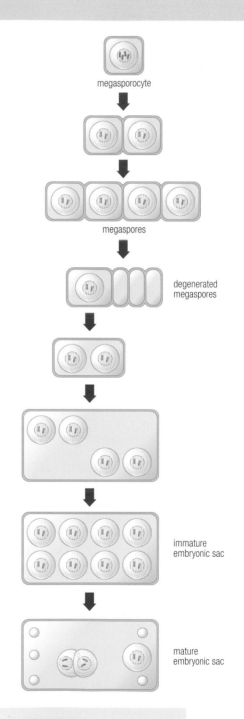

megasporocyte

megaspores

degenerated megaspores

immature embryonic sac

mature embryonic sac

In higher plants, female gametes are produced in an ovary, and male gametes are produced in anthers (both parts of the flower).

microsporocyte

microspores

grains of pollen

The process by which grains of pollen are formed (near right), and the formation of the embryonic sac in a higher plant (column at far right).

THE LIFE CYCLE OF PLANTS

The life cycle of plants is characterized by an alternation of generations between two very different types of individuals: the sporophyte and the gametophyte. The **sporophyte**, as the name indicates, is the phase that produces spores—in other words, the one that carries out asexual reproduction. When a spore germinates, it produces another type of plant, which is the **gametophyte**; as it grows and develops, it forms structures in which the gametes are produced. When these are used in sexual reproduction, they produce a new organism that is always **diploid** and once again a sporophyte. Thus, the cycle keeps alternating between generations of sporophytes and gametophytes.

Sporophytes and gametophytes can live separately (as in the case of ferns, algae, and mosses) or together (as with higher plants). In the latter case, it is very difficult to distinguish the two generations by eye.

ENVIRONMENTAL ADAPTATION

The oldest plants lived in an aquatic environment, since their male gametes needed this medium for movement toward the female gamete. In order to colonize a land environment, they had to change the system for moving the **male gametes**, continually adapting more effectively to the open air; the result is that the most highly evolved plants can live in the most arid regions, since the producer of male gametes travels inside a protective structure.

 Seeds are structures that encase and protect the plant embryo and keep it from drying out and allowing it to germinate only under the proper conditions.

Cacti are plants that have adapted to desert regions; they have also developed strategies for capturing the little bit of available water, storing it, and reducing it to a minimum loss through evaporation and damage by thirsty animals.

THE PARTS OF A FLOWER

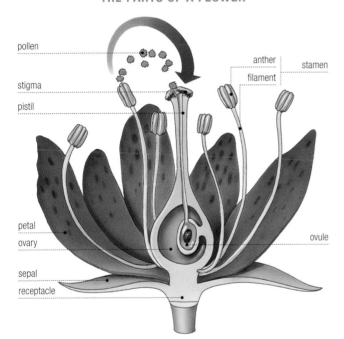

pollen
stigma
pistil
anther
filament
stamen
petal
ovary
sepal
receptacle
ovule

FLOWERS

This is the part that is responsible for **sexual reproduction** in higher plants. The male part consists of the **stamens**. They are where the spores are produced; inside them are the male gametes. The female part consists of **pistils**; these are bottle-shaped structures found in the **ovaries**, where the female gametes are formed. Pistils have an opening known as the **stigma**, through which grains of pollen enter.

Some species of plants produce female and male flowers separately. In some cases, the same plant has both types of flowers, and others have just one type.

Many flowers are attractive and emit pleasant aromas to attract insects to transport grains of pollen from one flower to another. Plants that use wind for pollinating have flowers that are much less striking in appearance.

SEEDS

A seed is the **embryo** of a plant. It is produced when a **grain of pollen** fertilizes an **ovule**, which then turns into a structure containing nutritive substances that are used in the subsequent development of the young plant, plus a protective covering. Gymnosperms (e.g., pine trees) produce seeds in the open or protected by some structure such as a cone; however, in angiosperms, the seeds are surrounded by a **fruit**, which can take on a wide variety of forms.

THE PARTS OF A SEED

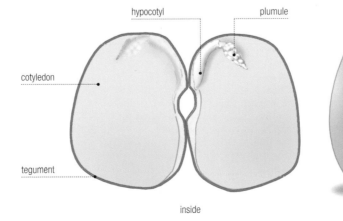

hypocotyl
plumule
cotyledon
microphyll
joint
tegument
inside
outside

SEXUAL REPRODUCTION IN ANIMALS

The most widespread means of reproduction in the animal world is sexual, that is, involving the union of two gametes, a spermatozoid and an ovule; when they are joined, they produce a zygote or an initial cell, from which the rest of the body develops. Only some lower groups (sponges, cnidarians, echinoderms, and some worms) are capable of reproducing asexually; all the others reproduce sexually.

FERTILIZATION

This is the instant when an **ovule** (the female gamete) is united with a **spermatozoid** (the male gamete). In nature, there are two basic forms of fertilization: either inside the female's body (**internal fertilization**), or outside (**external fertilization**). External fertilization is the norm with aquatic animals, such as crustaceans, fish, and others, where the male gametes can move through the water. Internal fertilization is used by land animals and some aquatic ones (such as sharks); it takes place through **copulation**, in which the male releases spermatozoids into the female's body.

The trout is an aquatic animal that lays its eggs in the water, and they develop in that environment. The photograph shows collecting eggs from a common trout for breeding alevins to restock rivers and lakes.

The union between two animals of the opposite sex is known as copulation.

With sea horses, the female deposits ovules inside the male's copulation sac, and it is the male that carries the eggs inside his body until they emerge as baby sea horses. This is an exceptional case in nature.

GAMETES

Some animals produce gametes that are exactly the same, and there is no distinction between male and female gametes; they are known as **isogametes**, and they exist in some groups of protozoa and invertebrates. However, most species exhibit sexual differentiation. Male gametes, known as **spermatozoids**, are smaller and have a long flagellum that propels them through a liquid medium. Female gametes, known as **ovules**, are somewhat larger and round, and they are incapable of movement through their environment.

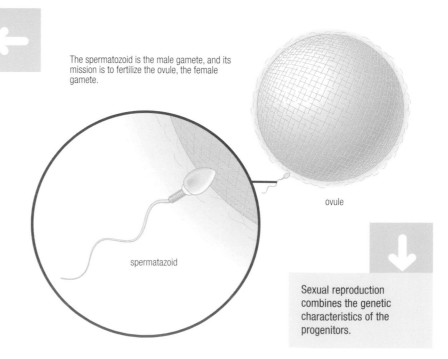

The spermatozoid is the male gamete, and its mission is to fertilize the ovule, the female gamete.

ovule

spermatazoid

Sexual reproduction combines the genetic characteristics of the progenitors.

THE PRODUCTION OF GAMETES

Animals have a couple of organs whose function is entirely for reproduction. The series of these organs is known as the **reproductive system**. There are a great variety of reproductive systems; the vertebrate system is the most widely known one, in which the male and the female exist separately. The main organs of the male are the two **testicles**; those of the female are the **ovaries**, of which there are, likewise, two. However, many invertebrates do not follow this pattern; rather, they may have more than two testicles or ovaries, and there are even individuals that are **hermaphroditic**—they possess an equal number of male and female organs at the same time.

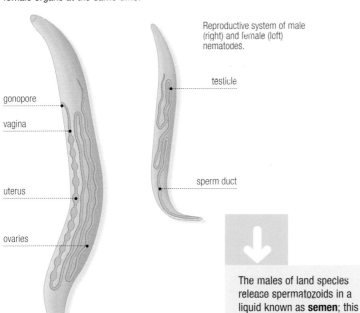

Reproductive system of male (right) and female (left) nematodes.

gonopore

vagina

uterus

ovaries

testicle

sperm duct

The males of land species release spermatozoids in a liquid known as **semen**; this allows the spermatozoids to migrate in search of an ovule.

Introduction

Life

Life on Earth

The Basis of Life

Biochemistry

Evolution and Genetics

Heredity and Genetics

Functions of Living Organisms

How Living Organisms Function

Relationships with the Outside World

Reproduction and Development

Classification of Living Beings

The Plant World

The Animal World

The Living World

Index

THE REPRODUCTIVE SYSTEM OF A HEN

ovary

oviduct

cloaca

THE RUT

Nearly all animals reproduce at a specific time of the year that is best suited for birth and growth of the offspring (favorable temperature, abundance of food, etc.). This reproductive period is referred to as the **rut**; generally it involves a major change in the animals' behavior, and some that usually live alone will live in pairs or a group for several days or weeks. In some species, there are actual courtship dances that take place before fertilization. In many cases, the males fight one another to gain access to a female. There are even some animals that change their physical appearance, generally heightening their color or developing noticeable structures on their bodies.

Many birds perform a very elaborate sexual ritual; the males wait for the females in clearings in the woods and attempt to attract them using very characteristic calls that sometimes sound like the clacking of castanets.

Humans and some primates observe no rutting season and can reproduce at any time during the year.

FROM EGG TO ADULT

In sexual reproduction, an ovule and a spermatozoid are united in the process known as fertilization; the result is a zygote. From this instant on, the zygote undergoes a series of transformations that make it resemble an adult more and more each day. This period is known as **embryonic development**; it is one of the most important periods of life, since it is when the organs necessary for life are formed.

IT ALL BEGINS WITH ONE CELL

When a spermatazoid unites with an ovule, they form a **diploid** cell known as a **zygote**, which develops into the creature's entire body. This first cell divides into two cells, which in turn divide into a body that has four cells, and the process of division continues to produce a large mass of cells. The initial stage of development is extremely important, since the first cells are **undifferentiated**; in other words, they can be used to create any part of the body. Later on, when the fetus has a certain number of cells, each of them will have a specific purpose; that is, they will make up a specific part of the body. If the fetus divides in two early on, twins are formed.

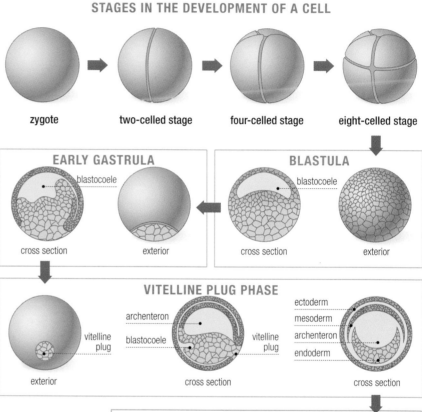

STAGES IN THE DEVELOPMENT OF A CELL

zygote · two-celled stage · four-celled stage · eight-celled stage

EARLY GASTRULA — blastocoele — cross section — exterior

BLASTULA — blastocoele — cross section — exterior

VITELLINE PLUG PHASE — archenteron — blastocoele — vitelline plug — exterior — cross section — ectoderm — mesoderm — archenteron — endoderm — vitelline plug — cross section

NEURAL TUBE PHASE — ectoderm — epidermis — neural canal — mesoderm — notochord — coelom — archenteron — endoderm

NEURAL FOLD PHASE — neural folds — notochord — mesoderm — coelum — ectoderm — archenteron — endoderm

Birth is the moment when a living being emerges from the inside of its mother's body or an egg to begin its separate life on the outside.

OVIPAROUS AND VIVIPAROUS ANIMALS

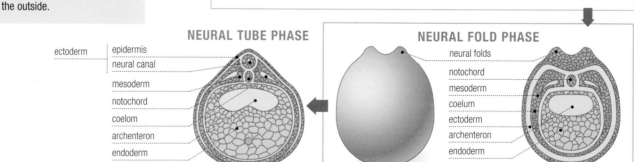

In some animal groups, such as the mammals, embryonic development takes place inside the mother's body, and the young emerge completely formed; these animals are said to be **viviparous**. When a female is pregnant, the **fetus** grows inside her body and gets necessary nutrients from the mother's body, which are transported in the blood through the **umbilical cord**. Other animal groups, such as the invertebrates, fish, amphibians, reptiles, and birds, lay eggs; they are **oviparous**. In this case, the fetus is closed inside a structure that contains all the necessary nutrients for its embryonic development.

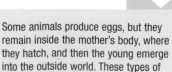
Some animals produce eggs, but they remain inside the mother's body, where they hatch, and then the young emerge into the outside world. These types of animals are said to be **oviparous**; examples include some snakes and insects.

DIRECT AND INDIRECT DEVELOPMENT

Animals can be born with a body similar to an adult's of the same species, and the only changes it undergoes in reaching adulthood involves organ growth and functionality. Other animals are born with a totally different appearance from the adults, and, in their development, they undergo a period of profound transformation known as **metamorphosis**, in which the body experiences a complete change. The former animals are said to be characterized by **direct development** and the latter by **indirect development.**

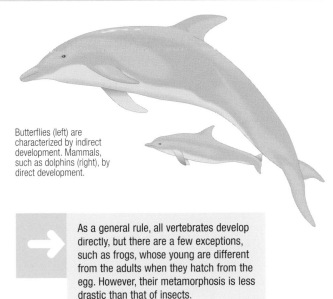
Butterflies (left) are characterized by indirect development. Mammals, such as dolphins (right), by direct development.

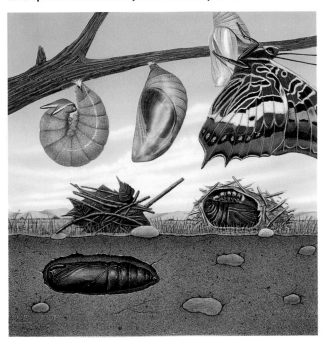

→ As a general rule, all vertebrates develop directly, but there are a few exceptions, such as frogs, whose young are different from the adults when they hatch from the egg. However, their metamorphosis is less drastic than that of insects.

GESTATION PERIOD

Between the time of **fertilization** and hatching or birth, there is a period that varies in duration according to species. Generally, the larger the animal, the longer it takes to develop, so the longer its gestation period.

MARSUPIALS

This group of animals is an exceptional case. Gestation begins inside the mother's uterus, as with all mammals, but after a while, when the fetus is still very underdeveloped, it emerges, crawls onto the mother's body, and climbs into a type of abdominal pouch known as a **marsupium**. This is where it completes its development. It gets its nourishment from milk provided by the mother's glands as long as it remains in the marsupium.

Marsupials are primitive mammals. The illustration shows a kangaroo with its offspring in the marsupium.

THE DURATION OF INCUBATION OR GESTATION

Species	Gestation or incubation
goose	1 month
beaver	40–70 days
salmon	2–3 months
polar bear	8 months

CLASSIFYING LIVING ORGANISMS

There are many different types of environments on Earth and that has resulted in a great variety of animals, plants, fungi, and microorganisms, all with their own characteristics that allow them to live in those environments. To understand nature, it is important to know its inhabitants; for that purpose, scientists have developed systems for classifying organisms.

TAXONOMY: THE SCIENCE OF CLASSIFYING

In olden times, organisms were classified simply on the basis of their similarities and differences, but it has been shown that grave errors were made, because organisms were placed in the same group since they resembled one another, but they were very different on the inside. For example, it was believed that **mushrooms** were a type of **plant**. Nowadays, the main characteristic by which living organisms are classified is the degree to which they are **related**.

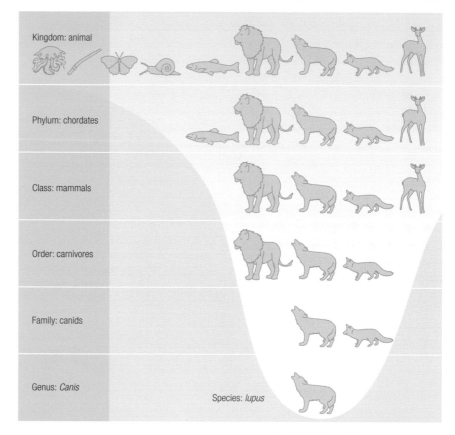

Kingdom: animal

Phylum: chordates

Class: mammals

Order: carnivores

Family: canids

Genus: *Canis*

Species: *lupus*

Many scientists group together, into a single kingdom known as **protists**, the eukaryotic organisms that do not exceed the cellular level of organization. This group includes the protozoa, one-celled algae, and one-celled fungi.

CLASSIFYING ORGANISMS INTO FIVE KINGDOMS

Kingdom	Organization	Nucleus	Nutrition
Moneran	one-celled	prokaryotic	autotrophic or heterotrophic
Protist	one-celled	eukaryotic	autotrophic or heterotrophic
Fungi	one-celled and multicelled	eukaryotic	heterotrophic
Plants	multicelled	eukaryotic	autotrophic
Animals	multicelled	eukaryotic	heterotrophic

Even though a dolphin looks more like a tuna fish than a camel at first glance, in fact, it is more closely related to the latter, since both share the characteristic of nursing their young. Both are mammals, even though they live in different environments.

HOMOLOGY AND ANALOGY

The outside appearance of an animal can confuse an observer, since environmental pressure can make two creatures from different groups resemble one another closely. **Homologous** structures are ones that have the same biological basis but whose external appearance is very different, such as a bird's wing and the front foot of a mole. **Analogous** structures are the opposite: They have the same function, but anatomically they are very different, as with the wings of a bird and an insect.

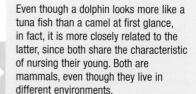

homologues

mole

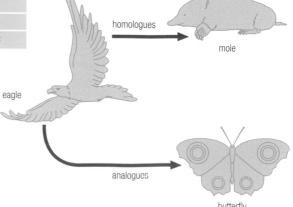

eagle

analogues

butterfly

THE BINOMIAL SYSTEM

All species of living beings can be identified with Latin scientific names that are made up of two parts (**binomial nomenclature**). The first is the genus to which the creature belongs; the second, the specific species. This scientific name was invented by the naturalist **Linnaeus** so that the species would not be confused with one another because of idiomatic differences. Scientists in all countries use this naming system. The common name is the one that people use in everyday language and that varies by country. Thus, the scientific name *Canis lupus* is *wolf* in English, *lobo* in Spanish, and *loup* in French.

Brown bear *(Ursus arctos).*

Sea horse
(Hippocampus guttulatus).

Beech tree *(Fagus sylvatica).*

Common snail
(Helix aspersa).

 Behavior is also considered in identifying living beings. For example, ornithologists commonly use the song or the means of flight in identifying different species of birds.

DIFFICULTIES IN CLASSIFICATION

It is very difficult to classify living beings, since the dividing lines between groups are not always very clear. For example, **protozoa** are organisms that function like animals, but their bodies are made up of just one cell, so they cannot be included in the animal kingdom, which, by definition, includes multicelled organisms. We find another instance of this in the **fungi**, whose general appearance largely suggests plants; however, their means of obtaining nourishment is so different—they are heterotrophs, while plants are autotrophs—that it would not be correct to place them in the same group. There are many similar examples of animals that have characteristics partway between two groups.

A branch of coral looks like an individual unto itself, but, if we get close enough, we can see that, in fact, it is a large calcareous skeleton on which many small polyps live.

MICROORGANISMS

As the name indicates, these are very tiny organisms, to the point that most of them cannot be seen with the naked eye and require the aid of a magnifying glass or a microscope. They are everywhere; the only place they cannot be found is in very extreme environmental conditions and places that have been disinfected and sterilized.

TYPES OF MICROORGANISMS

There are many different types of microorganisms, but most of them have a body composed of a single cell. Bacteria and cyanophyceae are **prokaryotes**; that is, their genetic material (DNA) floats freely around the cytoplasm. The other living beings, including many microorganisms, are **eukaryotes**—in other words, their DNA is located inside a cellular compartment known as a **nucleus**, which is separated from the protoplasm by a cellular membrane.

GROUPS OF MICROORGANISMS

Group	Cellular nucleus	Nourishment
Bacteria	prokaryotic	heterotrophic or autotrophic
One-celled fungi	eukaryotic	heterotrophic
One-celled algae	eukaryotic	autotrophic
Protozoa	eukaryotic	heterotrophic

Viruses are organisms located at the dividing line between living and inert matter, since they cannot perform life functions for themselves. They do not eat, they do not relate to one another, and they do not move. They only reproduce, and for that purpose they must get into a living cell and use its reproductive mechanisms.

EXAMPLES OF MICROORGANISMS

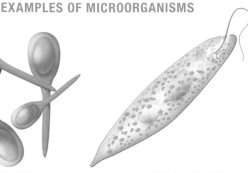

bacteria (bacillus)

one-celled algae (euglena)

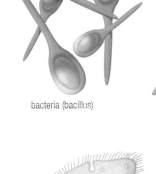

protozoa (paramecium)

mucilaginous fungi

THE BENEFITS OF MICROORGANISMS

Many microorganisms are very important for the **ecosystems**, and, without them, it would be impossible for nature to work properly. For example, in marine ecosystems, microorganisms are at the base of the **ecological pyramid**. To begin with, one-celled algae feed on the nutritive molecules in the water. Next, protozoa eat these bacteria, and they, in turn, serve as food for many small invertebrates. These invertebrates make up the diet of many birds, amphibians, and fish. As a result, if microorganisms did not exist, the rest of the animals could not exist.

THE NITROGEN CYCLE

dead plants

N_2 (atmospheric nitrogen)

NH_3 (ammonia)

nitrifying bacteria

microorganisms

Nitrogen is an indispensable element for living things, since it is used in making proteins. Only nitrifying bacteria are capable of fixing atmospheric nitrogen so that it can be used by plants.

MICROORGANISMS THAT CAUSE DISEASES

Many diseases are caused by parasitic microorganisms that get into the body one way or another and cause trouble. The microorganisms get in by various means—in the air that is breathed in, with water or food, through the skin, or through some other animal (for example, mosquitoes and ticks can introduce pathogenic microorganisms through their bites).

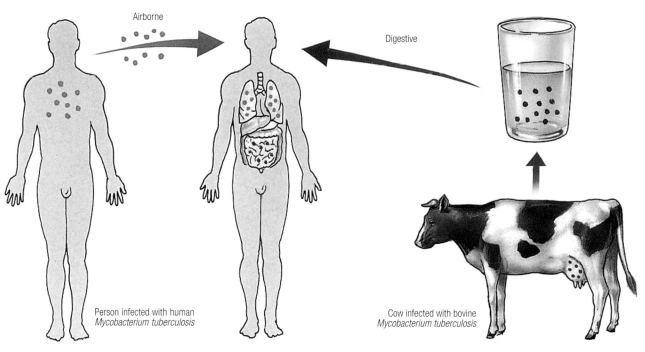

Airborne

Digestive

Person infected with human *Mycobacterium tuberculosis*

Cow infected with bovine *Mycobacterium tuberculosis*

Route of contagion by tuberculosis, caused by Koch's human or bovine bacillus (*Mycobacterium tuberculosis*)

Some of the most serious contagious illnesses are tetanus, diptheria, cholera, meningitis, scarlet fever, whooping cough, and leprosy.

STERILIZATION

This is a means of eliminating all the microorganisms on an instrument or object. Surgeons, dentists, and doctors generally sterilize their instruments before using them to avoid infecting the patient with microbes that may cause an infection.

There are other ways, the best known of which involves heat. Microbes generally cannot survive temperatures over 212°F (100°C), so boiling the objects or putting them into a hot oven kills them. Another way is to sterilize with alcohol or other chemical products such as chlorine, which are toxic to microbes.

The bodies of many living things are protected by an external covering—bark, in the case of trees, and skin, in animals. When this covering is torn it is very easy for microorganisms to get in through the wound and cause an infection.

SCHEMATIC CHART OF THE AIDS VIRUS

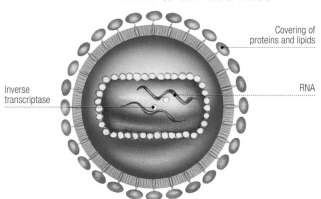

Covering of proteins and lipids

RNA

Inverse transcriptase

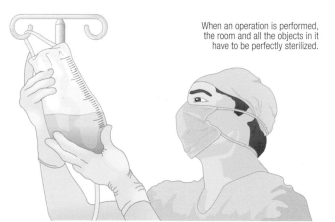

When an operation is performed, the room and all the objects in it have to be perfectly sterilized.

83

THE PLANT WORLD

Many diverse organisms are grouped together as plants, but the majority of them exhibit characteristics that are quite easy to recognize on sight: They are green, or at least part of their body is green, and they live immobile, attached to the substrate by their roots. Plants are autotrophs because of photosynthesis; in addition, their cells have chloroplasts and a hard cell wall made of cellulose.

ALL-TERRAIN ORGANISMS

Plants can be found in nearly every place on Earth, from the bottom of the ocean to the highest mountains. The main limiting factor in their development is the absence of **light**, so they are never found inside dark **caves** or in the deepest parts of the ocean where the sun's rays cannot penetrate. That is because they need light to live, since it is their source of energy. Another very important factor in plant growth is **water**; plants grow luxuriantly in moist areas and poorly in arid ones. **Temperature** also plays a role; cold outside temperatures do not permit growth or, at least, slow it down significantly, and, as a result, there is no vegetation in regions where there are glaciers or perpetual snows.

Nowadays, pollution also affects where plants grow; acid rain is the main factor in forest destruction.

INDISPENSABLE TO LIFE

Around two billion years ago, the Earth was a very different planet from what it is today. Its **atmosphere** was unbreathable, since the percentage of oxygen was very low. At that time, the first creatures capable of performing photosynthesis appeared: the algae known as **cyanophyceae**. Because of their continuous activity in capturing CO_2 and releasing O_2, (the **photosynthetic** reaction), eventually the cyanophyceae were able to change the composition of the atmosphere up to the 21% oxygen that it now contains. This made it possible for the rest of living beings that populate the globe to develop, since they need to breathe oxygen in order to live.

Plants are still the source of oxygen in the atmosphere. In fact, the Amazon jungle is called the planet's lungs, since it is a huge area filled with plants that purify CO_2 and generate O_2.

The Earth more than two billion years ago (left): only a few bacteria could live in the atmosphere with no oxygen. Today (right) the atmosphere makes life possible for an infinite variety of animals.

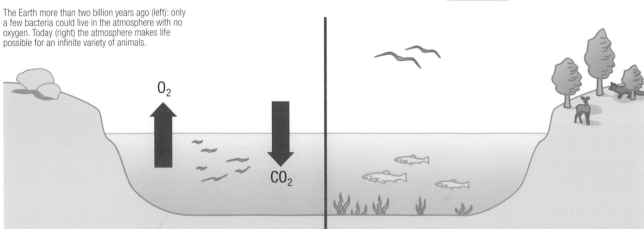

TYPES OF PLANTS

The plant world takes in a vast number of organisms. **Algae** are plants that live in aquatic ecosystems; their structure is simple, and they rarely have well-defined tissues. **Mosses** are plants that live in very moist areas; their structure is more complex than that of the algae, but they do not have any conductor vessels. **Ferns** live on land, but they still need lots of moisture for successful reproduction. They are equipped with conductor vessels. Finally, the **higher plants** (plants and grasses) are the ones that have succeeded in colonizing all the terrestrial ecosystems; they can live in truly arid regions, such as some African deserts, and they have very specialized conductor vessels.

Fungi are not plants, since they are heterotrophs; however, they were long thought to be plants because of their appearance. As a result, they are still included in many books on botany.

Ferns have simple conductors in their roots that absorb moisture.

Mosses are small plants that grow to form a green layer on the ground, rocks, and tree trunks.

Among the great variety of trees, the baobab is noteworthy; it is a tree of the African savanna that grows very slowly and flowers every seven years.

Flowers are nothing more than the reproductive organs of plants. Their shape, color, and scent serve to attract insects and assure fertilization.

GYMNOSPERMS AND ANGIOSPERMS

The higher plants are divided into two large groups: the gymnosperms and the angiosperms. The **gymnosperms** are characterized by their naked seeds, which may be protected only by scales (which make up cones), and they produce no flowers. Most of them are trees such as pines and birches. **Angiosperms** have their seeds inside a **fruit** that protects them and is produced from a **flower** after fertilization. These are the most abundant plants, and there are many different types: trees, bushes, and grasses.

Angiosperms are divided into the **dicotyledons**, which have two germinal leaves in the seed, and the **monocotyledons**, which have just one germinal leaf.

THE ANIMAL WORLD (INVERTEBRATES)

Animals are one-celled or multicelled and heterotrophic, with cells that have no cell wall. They can be divided into two large groups: **invertebrates**, which have no internal skeleton, and **vertebrates**, which have an internal skeleton. In this chapter, we will examine the invertebrates, which are extremely numerous and diverse, for they are found all over the planet.

INVERTEBRATES OF THE MARINE ECOSYSTEM

The oceans and seas are ecosystems that have a tremendous richness of invertebrates, and this is where the largest and most impressive ones live. Some of the most common groups are sponges, corals, jellyfish, snails, bivalves (clams, etc.), octopuses, squids, and crustaceans.

Invertebrates of the marine ecosystem.

Marine invertebrates are essential in the trophic chains and serve as prey for many fish, birds, and mammals, such as whales, which eat tiny crustaceans that make up the plankton.

FRESHWATER INVERTEBRATES

In the lakes and rivers throughout the world, there is a tremendous abundance of invertebrates. There are worms that live on the bottom and eat the organic material that accumulates on the floor. There are a multitude of microscopic crustaceans and terrestrial insect larvae that live in the water column; but these types of animals are found only where there is no strong current. In addition, some invertebrates live right on the surface of freshwater. This is not the case in marine ecosystems.

Insects are one class of arthropods, and it is believed that the number of species exceeds a million. Some insects are useful, since they produce products (such as honeybees and silkworms), pollinate plants, or purify substances; on the other hand, many of them are harmful, since they destroy crops, eat wood, and transmit diseases to humans.

Pollution is one of the most serious problems with freshwater ecosystems, and one of the main symptoms is a lack of invertebrates.

Freshwater invertebrates.

TERRESTRIAL INVERTEBRATES

On dry ground, the invertebrates that have been most successful are the **insects**, **arachnids**, and **myriapods** (centipedes and millipedes). Some of them can even live in the hottest desert, such as the Sahara. These invertebrates breathe through a series of tubes that penetrate into their body; these are known as trachea, and they deliver the oxygen from the air to the innermost tissues. There are also **mollusks**, such as snails and slugs.

Terrestrial invertebrates

Only a small group of **crustaceans**, the isopods, have succeeded in colonizing a terrestrial environment; they still live in areas of ample moisture.

On dry ground, the insects fulfill the same ecological purpose as the crustaceans in the marine environment.

THE UNDERGROUND WORLD

If we turn over a little soil with a shovel, we find a great number of invertebrates. Generally, these are animals that have no cuticle to protect them against desiccation, such as worms, which come to the surface only when it rains and environmental humidity is very high.

We also find invertebrates inside caves and rock crevices; generally, these creatures are light in color. In general, the invertebrates of the subterranean world have atrophied eyes because they cannot use them in such a dark environment.

There are many parasitic invertebrates, both internal (such as worms) and external (fleas and ticks); in addition to the harm they cause, they can also transmit diseases.

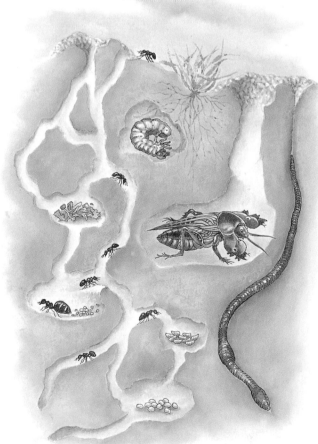

Invertebrates that live in the subsoil.

A great number of insects also live in the ground, such as the ants that construct a labyrinth of galleries and form large colonies.

THE ANIMAL WORLD (VERTEBRATES)

Vertebrate animals have an internal skeleton that serves as a support base for their muscles and allows them to move. In addition, the skeleton helps counteract the force of gravity and that allows them to grow to great size. Also, vertebrates have more complex nervous systems and a large brain that makes possible far more complex behavior.

This fish of the open waters has a light-colored lower body to make it harder to be seen by its predators in deeper water, and a dark back so it cannot be detected from above.

A fish's body shape may be an indication of where it lives: if the shape is very elongated, it lives in open waters; fish that live in the depths have a body flattened from top to bottom; and those that live on rocks and coral reefs have a body that is flattened from side to side.

Fish can also take on shapes and colors similar to the environment where they live so they can escape detection by predators.

FISH

Fish are the most primitive vertebrates, and all the other groups evolved from them. They live in oceans and rivers and breathe by means of **branchiae** or **gills**, which allow them to absorb oxygen from the water. They move through their liquid environment by moving their body, which propels them along, and their **fins**, which also help them change direction. All of them are **cold-blooded** (they cannot control their body temperature), their skin is covered with **scales**, and most of them are **oviparous**. There are two major groups: those with a **cartilaginous** skeleton, including rays and sharks, and those with a **bony** skeleton, which includes all the others (sardines, tuna, trout, etc.)

AMPHIBIANS

These are animals that evolved from fish and colonized a land environment; however, they still have certain characteristics that do not allow them to get too far away from the aquatic environment. Their **skin** is very delicate and permeable, and they produce a sort of mucus that helps protect them and keep them moist; that is not enough, though, and they still need to get wet from time to time. This group needs to **reproduce** in the water, where they lay eggs, since they are **oviparous**. The young have **branchia** or **gills** since they spend part of their life in the water. After a certain time, they experience a series of transformations (**metamorphoses**) in which the gills disappear, and they then begin using lungs as they turn into adults.

Some amphibians secrete toxic substances on their skin that keep them from being devoured by predators.

The word *amphibian* comes from the Greek *amphi* (which means *on both sides*) and *bios* (meaning *life* or *to live*). Thus, amphibious animals can live on both sides, in other words, on land and in the water.

REPTILES AND BIRDS

These are animals that are perfectly adapted to life on land, even though some of them have returned to the sea or to fresh water to live. Reptiles have a body surrounded by a **skin** that contains very hard, horn-like **scales** that keep them from drying out. Birds, which actually belong to a reptilian group, have exchanged their scales for **feathers**.

Both groups are **oviparous**, but reptiles generally abandon their eggs to their fate; birds, on the other hand, are more highly evolved and incubate their eggs and take care of the chicks once they hatch.

Another common characteristic is the presence of a **cloaca**, a cavity used in common by the excretory and reproductive systems and that opens to the outside.

The main difference between the two is that reptiles are **poikilothermic** (cold-blooded), and birds are **homeothermic** (warm-blooded).

Snakes are reptiles that have lost their legs. They move by undulating their body and with the help of their scales.

Dinosaurs were the largest living creatures that ever existed on our planet. They were reptiles of widely varying size and shape— terrestrial, airborne, and marine.

One of the major differences between reptiles (top; an iguana) and birds (above; storks) is that the former are covered with scales and the latter, with feathers.

MAMMALS

Mammals are the most highly evolved vertebrates. Their main characteristics are that their bodies are covered with **fur**, they are warm-blooded (**homeothermic**), and they give birth to live offspring (that is, they are **viviparous**), which they feed with **milk**, a nutritive substance produced in the **mammary glands** of the females.

They are **tetrapods**—in other words, they have four extremities, although, in some cases, they have eventually lost them. Most of them live on solid ground, but some have readapted to water (whales, dolphins, etc.), and others have acquired the ability to fly (bats).

Lions are carnivorous mammals, which means that they principally eat meat.

Deer are herbivorous mammals that mainly eat plants.

Bats are among the few mammals that can fly.

There is a small group of mammals that still lays eggs, the monotremes (ornithorynchus or duckbilled platypus and the echidna or spiny anteater). However, once the young emerge from the egg, they are suckled by the mother.

The white whale or beluga is a mammal adapted to water.

HUNTERS AND PREY

In the wild, there are many interactions between individuals of different animal species that share the same territory. One of the most important interactions is established between predators and their prey, since this is one of the main ways of controlling the balance of the ecosystem.

If predators did not exist, the prey animals would multiply to such an extent that their populations would quickly consume all the available food and they would all die of starvaton.

Predators evolve in tandem with their prey. The prey discover new ways to avoid the predators and that requires the predators to develop new skills.

WHAT ARE THE PREY ANIMALS?

These are the animals that are captured by others to serve as **food**. Generally, they are **herbivores** located farther down on the **ecological pyramids**. Like the rest of the animals, they have developed characteristics that allow them to flee from their enemies, but when they fall ill, get old, or are very young, they are not in top form, and they become vulnerable. It is usual for predators to use a minimum of **energy** in capturing their prey, so they always prefer animals that put up less resistance, rather than an adult in good shape.

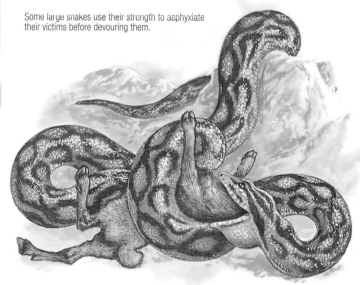
Some large snakes use their strength to asphyxiate their victims before devouring them.

HUNTERS

These are animals that capture other animals and eat them. They are **carnivores**, and they are at the top of the **ecological pyramid**. They are fewer in number than the herbivores, and they tend to be **territorial**, leaving signals on the ground that indicate to the rest of their kind that they must not come in to hunt, or they will risk a battle. These creatures commonly have features adapted to capturing their prey (claws, curved beaks, spurs, fangs, and so forth). Some of them have specialized by modifying their anatomy significantly, such as anteaters, which have a large, tubular snout that they stick into anthills to capture ants.

Scavengers are also carnivorous, but they do not hunt their prey; rather, they clean up the leftovers from predators that have finished eating.

In contrast to what humans sometimes do, animals never hunt more prey than they need; otherwise, they might annihilate a species they use for food.

The eagle, a carnivorous bird, has exceptional eyesight; it silently pounces on its prey from high in the sky.

DEFENSE AGAINST PREDATORS

Most animals flee from their predators or try to hide inside their den or nest. But others remain totally immobile to escape detection by the hunters. This is the technique used by newborn fawn. **Protective coloration** is another of the most commonly used characteristics that animals have to elude their enemies. Sometimes this involves **camouflage**, coloration that blends in with the surroundings. One example is the flounder, which can scarcely be seen from the ocean floor; another is the leaf insect, which looks like a plant. At other times coloration is very noticeable and is made up of warning colors (usually yellow and black or red and black), which indicate that the creature is poisonous. Some animals take advantage of these signs and take on those same colors to make any predators think that they are poisonous, even though they are not, and to discourage them from attacking. Many snakes do this.

Shells, as in turtles, and quills, as in porcupines and hedgehogs, are other means of defense against enemies.

Otters are carnivores that live on the banks of rivers and lakes; they mostly eat fish that they catch in the water.

Some animals become specialists in catching certain types of prey, such as the secretary bird, which goes after reptiles.

HUNTING TECHNIQUES

As prey animals have acquired increasingly effective ways of fleeing from their predators, the predators have had to perfect their hunting techniques in order to survive. Some, such as the leopard, have developed exceptional speed in running, even though the usual technique is to hide in vegetation and stealthily sneak up on the prey to capture it in one leap without expending too much energy in the attempt (**lying in wait** or **stalking**). Other animals, such as lions and wolves, have learned to hunt in groups, cooperating with one another to bring down an animal that they probably could not hunt alone. Owls, hedgehogs, and other nocturnal animals are guided mainly by hearing and smell.

Frogs and toads catch insects by shooting out their sticky tongue at them.

Even though the gnu is a larger animal than a lion, the latter is better armed with cunning, speed, claws, and fangs.

Many insectivorous birds such as swallows and martins capture their prey by flying with their beak open and using it like a funnel.

ECOSYSTEMS

An ecosystem is the unit made up by a biotope (the substrate plus the physical and chemical conditions) and a biocenosis (the living creatures who reside in the ecosystem). The living organisms of an ecosystem relate to one another and with the biotope. The conditions of the biotope are crucial for the lifestyle of the organisms that live there.

GREAT DIVERSITY

The type of soil, the altitude, the orientation to the sun, the climate, the plant and animal species that live there, and other factors can have an important influence on an ecosystem.

There are countless examples of **ecosystems**, since the entire planet, or a type of landscape such as a desert, an ocean, or a simple, short-lived puddle of rainwater, can be considered an ecosystem. Our planet also contains a great variety of **biotopes**: areas of rocks or sand, freshwater or salt water, frozen ground, and so forth. Each of these places is **colonized** by some type of living organisms, which in turn evolve to continually adapt more effectively to this environment. This tremendous diversity of environments and living organisms means that practically no two ecosystems are precisely the same.

The study of ecosystems is very complicated, since so many factors have to be considered. For example, in studying a forest, it must be borne in mind that part of the organic matter generated (leaves, trunks, etc.) can follow the river water as far as the sea (another ecosystem).

Every ecosystem has its own characteristics, from the Earth, which is a global ecosystem, to the branch of a tree, which is a mini-ecosystem.

ECOLOGICAL SUCCESSIONS

Indiscriminate cutting and intentional forest fires are real catastrophes, since the burned forests cannot perform their duty of renewing the air, and the ecosystem becomes inhospitable.

When an area is completely devastated by some type of natural disaster, such as the eruption of a volcano whose lava burns and covers up everything in its path, a **biotope** becomes entirely lifeless. Then the process of colonizing this territory begins. The first to arrive are some small pioneer species that do not require very hospitable conditions, such as the **lichens** that grow on volcanic rock with very little moisture. These lichens contribute slightly to the erosion of the rock so that some **plants** can put down roots. These roots combine with erosion from rainwater and wind to form a thin layer of **soil** in which higher plants such as bushes and trees can gain a foothold. With time, a forest will surely end up where there was nothing but bare rock to begin with.

As the ground becomes covered with vegetation, spaces appear in which animals can live, since shelter and food are present.

The process of colonizing the soil.

HABITAT AND NICHE

These are two important concepts in ecology. A **habitat** is a set of **biotopes** in which a certain plant or animal species can live. For example, the habitat of seals is a marine environment. A **niche** is the part of the habitat where the organism really lives and the resources that it offers; for example, in the case of the seal, the niche is the cold, polar waters and the surroundings, and the seal has become a specialist in catching medium-sized animals for food. Starfish also live in these waters, but they live on the bottom and feed on other creatures of the bottom and on dead animals that settle there. As a result, seals and starfish live in the same habitat, but they occupy different niches.

 Many animals occupy the same niche, but in separate ecosystems; for example, in a savanna, the predators are lions and leopards, but in a northern forest, the predators are wolves and foxes.

Seals and starfish share the same habitat but not the same niche.

Every ecosystem has its own equilibrium. If predators (such as owls) are exterminated by humans, their prey (such as rats) multiply to excess and destroy plants and crops; eventually they perish from starvation.

A VERY FRAGILE BALANCE

All ecosystems need a balance between the energy and the matter that goes into and out of the ecosystem; otherwise, the ecosystem decays. In order for this balance to prevail, all the pieces of the puzzle have to fit together properly. In nature, the pieces are the living organisms, and, if one species is lacking, there can be a very negative impact on the other species, which throws the system out of balance.

 Ecology is the science that studies ecosystems, the environment, and the living organisms that inhabit it, plus the relationships that exist among them all.

 If the predators are eliminated from an ecosystem, the herbivores multiply excessively and end up devouring all the plants.

HUMANS AND LIFE

At first, humans were more of an animal of nature and depended entirely on what it had to offer. With evolution, humans became more intelligent and learned to control nature. Little by little, people have modified ecosystems through their actions, and, nowadays, these ecosystems depend on humans as much as humans do on them.

EVOLUTION: HOW FAR WILL IT GO?

The planet Earth and all of nature have evolved to the extent that now they bear no resemblance to the way things were at first. However, this **evolution** has been slow, following its own pace, until humans began to take on more intelligence, some 100,000 years ago. Since that time, due to the changes that have occurred in the ecosystems, nature has changed a lot. The problem is that the changes have been so quick that many plants and animals have not had time to adapt to the new conditions, and numerous **species** have vanished. Starting a few decades ago, many people have become conscious of the problem, and they are trying to rectify it. However, it is often hard to do this, since what really needs to be changed is the modern lifestyle that exploits nature.

Just a few centuries ago, the area where the city of Seattle is located was the site of beautiful tree-covered hills. In constructing the city, those hills were flattened.

This landscape (Rocky Mountains, Canada) has been the same for millions of years.

 The future of the planet is in our hands.

CLIMATIC CHANGE

One of the effects of human intervention in nature that can have a tremendous impact on life on our planet is the modification of general climatic conditions—in other words, **climatic change**.

Major alterations have already been observed, such as a general warming of the planet by a few degrees in the course of a century, a higher occurrence of natural disasters such as **floods** and **hurricanes**. However, since the period of time under study is very short in comparison to the age of the Earth, we still do not know if this is a natural cycle or the result of human activity. In any case, the **pollution** that is causing these changes is harmful by itself, so it is a good idea to avoid a possible change in our climate.

Climatic change has caused great disasters: ongoing drought in some areas, torrential rains in others, hurricanes in new places . . .

THE OCCUPATION OF TERRITORY

In the last hundred years, humans have done as much construction as in the preceding several thousand years, and cities, highways, dams, canals, and many other public works have taken over ground space; the results have been landscape modification, disappearance of plant mass, and the flight or extinction of many animals. Even though this invasion may be necessary, it has almost never respected the environment. If the dangers of unchecked urbanization are not taken into account, the damage to life on Earth, including humans, may become irreversible in a very few decades.

FOSSIL FUELS

Lush forests have existed for millions of years, absorbing CO_2 from the atmosphere and incorporating it into organic plant material. This reaction was accompanied by the formation of O_2, so that the atmosphere has reached its current composition, which allows present organisms to live. Many of these forests died off and were covered with layers of sediment; in the absence of oxygen, the organic material was transformed into **coal** and **petroleum**. The CO_2 stored in these materials was thus buried under ground. When humans extract this petroleum and burn it, the CO_2 is released once again into the **atmosphere**, and the combustion process also uses up O_2.

How much longer will the automobile remain a symbol of the progress of our time?

In future years, rampant industrialization is going to exact a toll that humans may not be in a position to pay.

The increase of CO_2 in the atmosphere is one of the factors that contribute to the greenhouse effect; it produces a layer in the atmosphere that interferes with heat release, which increases the planet's temperature.